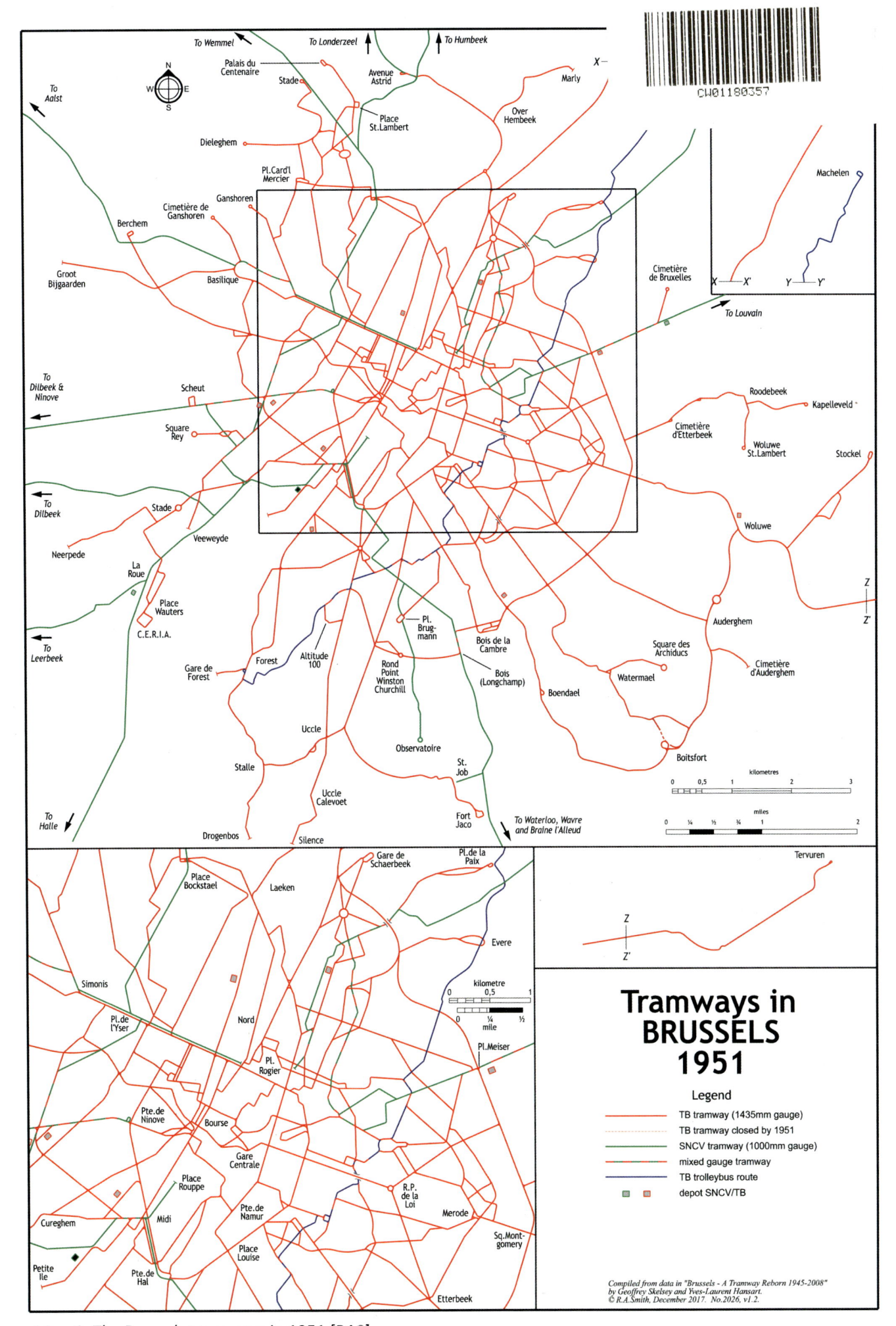

Map 1. The Brussels tramways in 1951 [RAS]

A classic scene on the approach to the Midi subways, with a four-axle PCC drawing up to a typical motor-and-trailer set before entering the tunnel down the four-track Jamar ramp. [MJR]

BELGIUM UNDERGROUND: Pre-metro and Metro, 1957 – 2017

Geoffrey Skelsey and Yves-Laurent Hansart

ISBN: 978-0-948106-56-9

Featuring photographs by Michael J. Russell and by the authors, and others as credited.
New maps and diagrams by Roger Smith.

Well into the 1960s the common type of car on much of the network was still the traditional motor-and-trailer set, like this one seen at Place Rogier in 1963. No 1248 was built in 1934, but modified after the War. The set is followed by one of the fine 1935-bogie cars of type 5000. The surroundings here would soon be excavated for construction of pre-metro lines 2 and 3 (and in 2018 are being altered again).

Dedicated to the memory of M. Guy Cudell (1917 – 1999)
Politician, Mayor of Saint-Josse-ten-Noode,
President of the Administrative Council of the STIB, and determined advocate of the tramways.

πάντα χωρεῖ καὶ οὐδὲν μένει
"Everything changes and nothing stands still".
(Heraclitus of Ephesus, c.535 – 475 B.C.)

Published by the Light Rail Transit Association
138 Radnor Avenue, Welling DA16 2BY

www.lrta.org

Printed by:
Latimer Trend & Company Ltd, Estover Road, Plymouth, Devon PL6 7PY

Copyright © 2019 Geoffrey Skelsey and Yves-Laurent Hansart
The moral right of the authors has been asserted.
All rights reserved.

Without limiting the rights under copyright as reserved above, no part of this publication may be reproduced, stored in, or introduced into a retrieval system, or transmitted in any form or by any means (electronic, mechanical, photocopied, recorded, or otherwise) without the prior written permission of both the copyright owners and the above publisher of this book.

FRONT COVER: The essence of the pre-metro: a 7900-type 8-axle PCC emerges from the temporary portal south of Place Louise, opened in 1985. North of here the tunnel had been completed stage by stage since 1970, initially for tram operation, but in 1988 the whole route was converted to Metro. This evolution is the subject of this book.

Contents

Authors' Preface 6
1. Introduction 8
Outline Brussels history 9
Definitions 10

2. Origins of the underground tramway 11
2.1. A tram subway round-up: including Boston, Philadelphia, London, Los Angeles, Rochester, Newark, Washington DC, Fort Worth, Cincinnati, Stockholm, Sydney, Oslo, Marseille, Budapest, and San Francisco 12
2.2 Germany's post-war plans 16

3. 'The Second Level': Tramway Planning in Europe after 1945 17
3.1 Leeds, Glasgow 18
3.2 Case studies in Continental Europe: 18
3.21 Hannover 19
3.22 Köln (Cologne) 20
3.23 Wien (Vienna) 21
3.24 Stuttgart 22

4. The Belgian and Brussels context from 1945 24
4.1 Introduction 25
4.2 Factors leading to investment in public transport infrastructure 25
4.21 New commercial and government quarters 26
4.22 Completion of the Junction railway 27
4.23 'Expo '58 28
4.24 Hospitals, Offices, and Universities 30
4.3 The Midi Subways 30
4.4 Wider Subway Plans from 1964 35

5. The Pre-Metro arrives in Brussels 37
5.1 Introduction 38
5.2 National and Regional Studies: 'PTU' and 'SSE' 42
5.3 Why pre-metro? 42
5.4 The achievement 46
5.41 Line 1: East-West tunnel 46
5.42 Line 2: The Ceinture (Inner Circle) Tunnel 49
5.43 Line 3: North-South Axis and extensions 52
5.44 Line 5: Eastern Orbital 57
5.45 Minor Underpasses 59
5.5 Pre-Metro Operation and Rolling Stock 60
5.51 Operation 60
5.52 Rolling stock since 1957 61

6. Conversion to Metro from 1976 65
6.1 Line 1 (1976) 67
6.2 Line 2 (1988) 68
6.3 Type 10,000: a lost opportunity 72
6.4 Line 3 plans: the 2001 studies 72

7. Pre-metro in other Belgian Cities 77
7.1 Antwerpen 78
7.11 The City 78
7.12 National Plans and initial response 79
7.13 Pre-metro as built, 1975 – 2017 81
7.14 Surface extensions and service list 86
7.15 Rolling stock 86
7.16 Conclusion 88
7.2 Gent 89
7.3 Charleroi 90
7.31 History 92
7.32 Pre-metro 92
7.33 Rolling stock 93
7.4 Liège 94

8. The STIB Metro develops **95**
A Brussels Metro in 1926? 99
8.1 A 'Designer Metro' 99
8.2 Trains 104
8.3 Routes 104
8.4 Present service pattern 108
8.5 The Future 108

9. Line 3: Twenty-first century questions
The decision to create a third Metro line between Albert and Bordet by the 2020s 109

10. Brussels Tramways in 2017 ....**113**
10.1 A long decline is reversed 115
10.2 Rolling stock update 115
10.3 Network 116
10.4 A look ahead 118
10.5 Tramway services in 2017 120

11. Conclusion
Pre-metro: delusion or salvation? 121

12. Acknowledgements and Bibliography **126**

List of maps and diagrams
Original cartography by Roger Smith marked 'RAS'.
Unless otherwise stated maps relate to Brussels.

1. The Brussels tram network in 1951. [RAS] .... [front endpaper] 1
2. Philadelphia tram subways in 1955 13
3. Fitzpayne plans for Glasgow, 1948 18
4. Hannover tram subways proposed in 1968 19
5. Köln 1968 plans 21
6. Wien 1960 plans 22
7. Brussels Expo 1958: tramway facilities 28
8. Midi subways 1957 31
9. Brussels tramways in 1969 [RAS] 36
10. Tram Subway proposals, 1960-63 [RAS] 39

11. Central area proposals, 1960-63 [RAS] 40
12. Tram traffic density diagram 1963-4 41
13. Pre-metro developments, 1969-70 [RAS] 44
14. Central area pre-metro, 1969-70 [RAS] 45
15. Tramways and metro around Gare du Midi from 1993 [RAS] 55
16. 'Grande Ceinture': pre-metro and metro connections at Montgomery [RAS] 58
17. The pre-metro in 1972 as shown by STIB 60
18. Pre-metro and metro, 1972-79 [RAS] 64
19. Central area pre-metro and metro, 1972-79 [RAS] 66
20. Light rail proposals published in 1979, for completion by 1990 [RAS] 69

21. Pre-metro and metro, 1980-90 [RAS] 71
22. Central area pre-metro and metro, 1980-90 [RAS] 73
23. Antwerpen 1962 plans 79
24. Antwerpen 1967 plans 80
25. Trams and light rail around Antwerpen 2017 [RAS] 84, 85
26. Subway proposals for Gent 1968 89
27. Initial proposals for light rail around Charleroi 91
28. 1926 plans for a metro in Brussels 96
29. Brussels pre-metro and metro, 1991-94 [RAS] 97
30. Central area pre-metro and metro, 1991-94 [RAS] 98

31. The STIB's line diagram for 2000 showing metro and pre-metro 101
32. Metro and pre-metro in February 2018 [RAS] 102
33. Metro and pre-metro February 2018, inner area [RAS] 103
34. 1962-3 STIB service diagram 114

Authors' Preface

The authors have known the Brussels tram subways almost throughout their lives. Geoffrey Skelsey first saw the Midi tunnels early in 1959, as a fifteen-year-old at school in Brugge. Yves-Laurent Hansart, a native Bruxellois, was born and brought up with trams passing his door and went daily to school on a 7000-type. Considering the hopes of the early pre-metro era it is easy to regret the course the story has taken, but on the other hand the resurgence of the STIB tramway since its low point in the 1980s is satisfying: tram traffic has roughly doubled over the last decade. Meanwhile in Antwerpen, where development took a different course as we shall see, the tramway has also flourished, but in a more conventional and consistent manner.

We stress that this book is not authorized by the STIB or other undertakings, and represents only the authors' opinions. Unless otherwise stated photographs are by the authors or from their collections. Illustrations marked 'MJR' are by Michael J. Russell. The provenance of some pictures is unknown despite inquiries and we will be glad to add full attribution in future. We are particularly grateful to Roger Smith, master map-maker, for the new plans which embellish the book.

We are very grateful to Mike Russell for painstakingly reviewing the text during preparation and saving us from many errors whilst adding important extra details from his own extensive knowledge of the Brussels and other systems. We have also been greatly helped by Martin Dibbs, David Holt, Tim Figures, and Eric Smith who generously reviewed the text at various stages. Any mistakes which remain are our own.

The primary subject of our book is a technology and operational practice as it developed in Belgium in the 1960s, to provide off-street rail transport in an affordable and realisable manner. This involved the phased construction of underground railways, to be initially operated as part of a conventional tramway system and brought into use stage by stage as work was completed; ideally, related to this was improvement of surface routes beyond the tunnelled sections to provide an equivalent quality of service. Eventually the completed tunnels and surface extensions could be converted to heavy underground railway ('Metro') operation. To make this possible the standards of the subway sections would need to anticipate ultimate Metro operation in terms of dimensions and other factors. As we shall see, the national government encouraged major Belgian cities to follow this path, and initially three did. Only in Brussels, however, has the evolutionary phase been fully followed through, with the conversion of most of the former tramway tunnels to conventional 'heavy' underground railways of the 'tube' or 'U-Bahn' type. For the preliminary tramway phase the Belgians later coined the term 'pre-metro', and we have used this generally throughout.

Our book starts, not in Belgium but with a survey of the tram subway concept as it developed after the late nineteenth century. This is followed by case studies of other European operators: these introductory sections highlight not only contrasts with the course taken in Belgium, but also Brussels' role as a pioneer in post-war tramway theory. We discuss later the case for and against the conversion policy, which Germany –in particular—did not follow.

We go on to review the political and economic situation in Belgium and Brussels in the 1950s which made tram subways possible; and then a detailed survey of their construction. We look at similar plans in Antwerpen and Charleroi, under the national Belgian proposals for city transport, although the predominance of the far larger Brussels installations will inevitably occupy most of our story.

We next examine development of the Brussels Metro since the first line was opened in 1976, and then discuss the factors explaining the progressive conversion of the city's pre-metro lines, including current plans for the trunk north-south line. Our studies ended early in 2018, and for subsequent developments the LRTA's monthly magazine 'Tramways and Urban Transit' provides regular reports.

Language is always an issue in Belgium, and we have had to be apologetically pragmatic in our usages. Expression in all the national languages is cumbersome, and in this book, mostly because they are more familiar to English readers, the French forms of place names are used within Brussels, thus for example 'Midi' (not 'Zuid'). When referring to places elsewhere in Belgium the locally-accepted form is adopted, such as 'Liège', and 'Antwerpen'. As an exception we have, however, adopted the English names 'Belgium', 'Brussels', 'Germany', 'Austria' etc. In most sections place names are as used by the inhabitants.

In keeping with practice in Belgium since 1820 we also use metric measures throughout: to add imperial units would produce awkward figures which most readers would probably find unhelpful.

To explain some of the transport terminology, the 'Société des Transports Intercommunaux de Bruxelles' (STIB) is the current tramway, bus, and metro operator in Brussels. The 'Société Nationale des Chemins de Fer Vicinaux' (SNCV or Vicinal) was the separate metre-gauge operator in many towns and cities, including Brussels, Antwerpen, Gent, and Charleroi. We have used the French-language acronyms for these bodies, which are explained later. 'SNCB' denotes the Belgian national rail operator, Société Nationale des Chemins de Fer Belges.

Amongst other Brussels usages 'Pentagon' refers to the five-sided shape of the inner city, within the line of the 'Petite Ceinture' ('little belt') which follows the line of the former city walls.

Some material in this book was first explored in our previous work Brussels: A Tramway Reborn (LRTA 2008), but is expanded and reconsidered here in a new context. Similarly, some historic photographs have been reused as the best available illustrations of particular topics, but the majority have not been used before and many are published for the first time here.

Geoffrey Skelsey
Yves-Laurent Hansart
Cambridge and Brussels
March 2017 to February 2018
geoff.skelsey@icloud.com

1. Introduction

Beneath today's heavy traffic and thronging crowds ghosts of a more dignified old Brussels can still be found. This is Place Louise before 1914, with a Chemins de Fer Economiques semi-open tram heading for Rue Royale in the foreground, on tracks still in use today. Crossing the view from left to right are the tracks of the Petite Ceinture, Metro since 1988. The shape of the Square is still discernible today, and a few of these buildings survive (see page 51).

1.1 There was a spirit of optimism in Belgium in the autumn of 1957. The long agony of the war and occupation, and the bitterness which followed, were fading. Prosperity was returning. Shops and cinemas were crowded. In the capital, the city centre at times seemed like a huge building site, as work proceeded on road and other improvements associated with the impending international exhibition, due to open the following spring. Of course, similar renewal was apparent in most European cities, but in Brussels – and later elsewhere in Belgium- these included the modernisation of the tramway networks, in contrast to Great Britain where the last cars ran in Liverpool that very year and trams were clearly in final decline, as they also seemed to be in France. At that time the Brussels standard gauge network alone amounted to over 240 route-km, and there were nearly 1000 motor trams. A few lines had been converted to buses, but they were still a minor element. As in some other European cities a solitary trolleybus line was never repeated and would soon close.

The completion just before Christmas 1957 of some 750 metres of tramway subway near the Midi station in Brussels will not seem a major event by the standards of what has followed, but in the circumstances of the time it attracted extensive attention, not least because the project came to fulfilment after nearly ten years of actual construction, and was a major feat of civil engineering.

Following the success of the Midi venture in improving the flow of both trams and motor vehicles those tunnels became the model on which future plans in Brussels were based, and in particular led to the concept of the 'pre-metro' which has been shared across the Continent, notably in Germany, as well as elsewhere in Belgium. As we near the end of the second decade of the twenty-first century, sixty years after the first Brussels tunnels were opened, the limited future of one of those remaining in tram operation there is becoming sadly clear. Within a decade or so from now these tunnels, like those of two of the earlier pre-metro lines, will have been partly adapted to form part of the heavy Metro system, an explicable development given the increase in traffic but one which contrasts with practice in some other parts of the world. Meanwhile in Antwerpen the pre-metro system continues to grow in line with the original tram-operated concept. Modest extensions to the small Charleroi system are possible, but in Gent subways were wholly eschewed. The plans for subways in Liège were never completed.

It is therefore timely to survey the history of the Brussels tram subways and to see how they developed over time: in one sense they are the victims of their own success, because they proved the need for, and the beneficial consequences of, reliable fixed-track transport along the routes they served. The contrasting developments elsewhere in Belgium are also illuminating.

Once such impending changes as these might have been sounding the death knell for the remaining tramways in Brussels, and this sometimes seemed possible in the city. There is little chance of it happening now as the tramways have been recast and largely re-equipped, providing with new lines important high-capacity feeders to the Metro and creating cross-links where full Metro capacity is unjustified. Renewed and reorganised the Brussels trams will be playing a different but vital role for many years to come, as they will in the three other relevant Belgian cities.

One thing above all emerges from this story: the benefits of foresight and planning, which mean that the transport proposals embarked upon over sixty years ago were largely followed through, to lay the foundations for today's public transport systems.

1.2 History

Because it affects our story we need briefly to consider the history of street tramways in Brussels, and in particular the origins of the present operating body, '*La STIB*', which planned and built the original pre-metro. (Outlines of the other Belgian cities are contained in their respective sections).

Street tramway operation in Brussels began in the 1860s under the aegis of numerous competing entities. Gradually order emerged in the form of two standard gauge operators, *Les Tramways Bruxellois* established in 1875, and *Les Chemins de Fer Economiques* in 1889. These bodies eventually merged and their routes were the basis of much of the network we see today. Electric traction began in 1894, with conduit operation in the central area to avoid the offence of overhead wires. This lasted in part until 1942. A fundamental administrative reorganization occurred at the start of 1946, when the last concessionary period of the *Tramways Bruxellois* company expired. In immediately post-war circumstances a provisional body the *Comité provisoire de gestion des Transports Urbains de l'Agglomeration Bruxelloise* (TUAB) was established to manage the undertaking pending consideration

Metre gauge at Place Rogier: this is the new Vicinal terminus in 1958, with the 'Martini' building going up in the background on the site of the old Gare du Nord. In the distance standard gauge tracks and a STIB tram. [Courtesy Tony Percival]

of a permanent administrative structure. The TUAB would last until the end of 1953 when it was succeeded by a new organisation set up by legislation in June 1953, *La Société des Transports Intercommunaux de Bruxelles* (STIB). This was initially a public-private partnership involving on the one hand the Belgian State, the Province of Brabant, the City of Brussels and the other communes of the Brussels district; and on the other the *Tramways Bruxellois S.A.*, which provided half the capital. Although modified later, the STIB continues in existence after 65 years, a striking contrast to the structural reorganizations of transport in London over the same period.

Brussels was unusual amongst European capitals in having two quite distinct (but interlocked) tramway undertakings, which furthermore operated on different track gauges. The mainly metre-gauge *Société Nationale des Chemins de Fer Vicinaux*, SNCV or 'Vicinal', was originally formed in 1885 as a quasi-governmental supervisory body overseeing concessionary operation of publicly-owned rural tramways. In time the organisation moved far beyond this original concept, notably by providing interurban and suburban passenger services, and it also took on direct operation itself. The SNCV had introduced electric operation around Brussels from 1894, and for legal reasons had a monopoly of operation on the principal 'national' roads; it also operated several exclusively urban services within Brussels, including a short tramway subway, as we shall see.

1.3 Definitions

To avoid repeated explanations we need briefly to consider some of the terms we use in our title and throughout the book.

The evolving nomenclature surrounding the tram subway is significant. The concept began in North America and borrowed local usage ('subway' being of course an American term for what was called elsewhere an 'underground', 'tube', 'U-Bahn', or 'Métro'). Interestingly the usage was borrowed by the London County Council for their Kingsway tunnel (as well as in Glasgow), but in England became the customary term for a pedestrian underpass. The increasing disfavour into which the street railway or tramway fell in the course of the last century led to a series of successive coinages: in the United States the term 'subway-surface operation' was adopted in place of 'street car subway'. In Germany a 'U-Strassenbahn' became a 'U-Bahn' and sometimes (as in Frankfurt) a 'Stadtbahn', although that also has a different meaning in, for instance, Wien. In Switzerland the term 'Tiefbahn' ('deep railway') was briefly adopted. The Brussels project began with the factual usage 'tramway souterrain' ('underground tram') but this morphed into 'semi-métro' and then 'pre-métro', attempts both to present the development as a radical new mode, and to avoid any idea that it was a temporary or second-best solution. In Charleroi the pleasant term 'Métro-Leger' (light metro) has been used, as it has been recently in Brussels.

In some cases the interim 'pre-metro' phase has lasted for over thirty years, and in others may well do for ever.

Although fifty years ago the evolution of underground tramway into underground railway, as defined in Belgium, was thought to offer an example to other operators this has not in fact turned out to be the case, and the Brussels installation remains unique in the way it has developed. Elsewhere in Belgium the pre-metro is still a current mode.

This is its story.

Before we consider the planning and execution of the Belgian tram subways we review the development of the concept around the world from its origins in the nineteenth century. We shall identify a number of themes with which to compare the Belgian achievement. Although there was a handful of early European applications, the technology flowered in the United States before activity moved back to the Old World in the 1950s, and if Brussels was amongst the pioneers in the post-war trend it was unexpectedly in Great Britain that aspects of it were first defined.

RIGHT: The original portal at the Boylston Street end of the Tremont Street subway in Boston set the pattern for tram tunnel ramps to the present day, and would be recognisable in Belgium now.

2. Origins of the Underground Tramway

2.1. A subway round-up

Until the 1950s tram tunnels were built to cater for distinct local needs, such as avoidance of surface traffic congestion and preventing disturbance to local amenities. There was, as yet, no thought of offering a comprehensive network like those of conventional underground railways, nor was there yet any related attempt to improve performance on associated surface sections by segregation or by traffic priority for trams: such measures lay far in the future.

Looking first at the more important applications, the history of the tram subway began in **Boston**, Massachusetts in September 1897. The Tremont Street line was explicitly intended to relieve traffic congestion in the centre of a city with (by North American standards) relatively narrow and irregular streets, and provided access for a range of existing street car routes. First planned in 1887 work on the new tunnels began in March 1895, and they were built partly below public gardens and partly by cut-and-cover construction below successive streets. Although short the installation was relatively complex with three separate portals, a grade-separated running junction, a section with divided tracks beneath adjacent streets, and two underground turning circles. There were five stations. The Tremont Street subway was subsequently much altered and still forms a central part of the 'Green Line' group of light rapid transit services, with some original features still apparent.

With Boston **Philadelphia** offers the prime surviving example of an early tram subway designed as part of a coherent plan to bring suburban street cars into the central city. It was opened in December 1905 with its tracks either side of those of the Market Street heavy rail subway, and originally extended between 22nd and 14th Streets, later with a loop beneath the City Hall. It was extended further west in 1955 beyond the Schuylkill River, diverging from the Market Street subway beyond 30th Street and then splitting into two branches. There are now eight intermediate subway stations and the tunnel carries cars of five different suburban routes.

The solitary British application, the well-remembered Kingsway Tramway Subway in **London**, was sadly only a small part of what was planned to be a network of tramcar tunnels serving the City and West End. Its principal motivation, built as it was below a broad new thoroughfare, was not the relief of traffic congestion but the need to link the northern and southern portions of the growing electric tramway system without incurring implacable parliamentary and local authority opposition to street tramways in the district. Authorised in 1902 public service over the northern part as far as Aldwych was inaugurated in February 1906 (the tramways along the Victoria Embankment, to which the subway was eventually linked, did not yet exist). The complete line was opened in April 1908, with a handsome archway leading onto the Embankment adjacent to Waterloo Bridge. For financial and technical reasons the line was initially built to accommodate single-deck vehicles only, but this was found to be unduly restrictive and it was closed between 1929 and 1931 for enlargement to admit double-deck trams. There

The Philadelphia street car subways were somewhat utilitarian: extended below the Schuylkill River in 1955 the subway served the main long-distance railway station at 30th Street, seen here with an inbound PCC car. The tracks of the heavy rapid-transit lines lie behind the fence on the right.

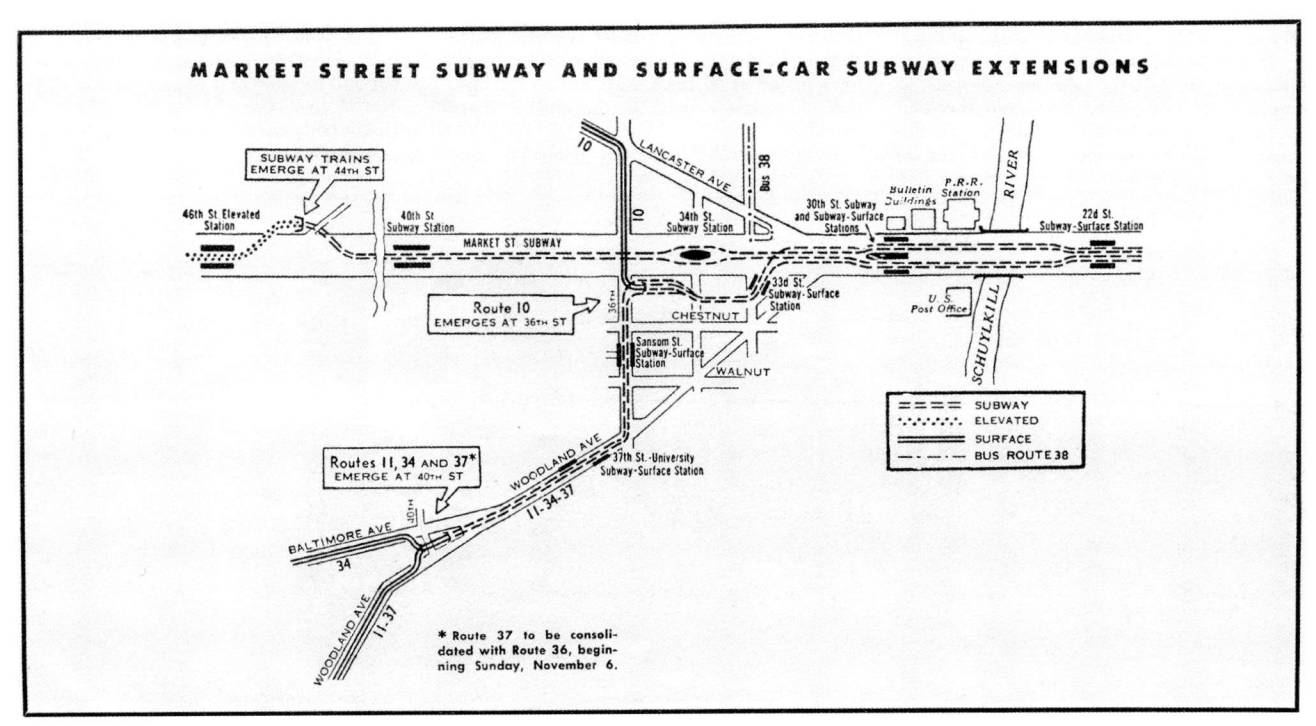

Map 2. Issued by the transit operator in 1955 this diagram neatly illustrates the new Philadelphia 'subway' and 'surface' routes west of 22nd Street, with the divergences leading to a range of street car lines, some still running today.

were two intermediate stations, at Holborn Station and Aldwych, with narrow island platforms reached by staircases located in the centre of the street (a third station, at the Strand, was omitted as an economy). The subway was a classic application, handling through services between a range of destinations north and south of the Thames. After closure in 1952 the tunnel stood idle until the southern portion was converted into a one-way road underpass for light traffic. The northern section, including Holborn station, remains in place unused today.

Perhaps inspired by the success of the Kingsway Subway there were several abortive plans for tramway tunnels, including in Birmingham (whose General Manager had been on a study tour of the United States), and Liverpool in Great Britain.

Successive proposals for **Toronto** are especially interesting as they originally envisaged a classic surface/subway installation which eventually took shape as a heavy rail metro, Canada's first. In 1909, when the city had a population of about 400,000, there was a British proposal for a street car tunnel below the main north-south thoroughfare, Yonge Street, with a branch to Broadview/Danforth, and new surface lines beyond the portals. Similar projects followed in 1910 and 1911. Funding required approval by referendum which was adverse, and the plans came to nothing.

With the entire street railway system in unified public control under the Toronto Transportation Commission from 1921 planning turned to re-equipment and modernisation, and it was not until the 1940s that subway proposals were revived. By then the city's population exceeded 900,000 and street car traffic had doubled over the preceding thirty years. Congestion and delay were rife, not helped by relatively narrow downtown streets. In 1942, as Canada looked optimistically towards post-war planning, the TTC proposed street car subways below Bay Street, a north-south thoroughfare west of Yonge Street, with two branches diverging north of Bloor Street. Also proposed was an east-west line below or near to Adelaide Street in the city centre, between Bathurst Street and River Street, with twin portals at each end of the subway.

Suburban street car routes, including some new construction, would have fed into the tunnels, providing a wide spread of travel opportunities. However reassessments of traffic growth, and observation of the new heavy rail subways in Chicago, brought a change of plan and preference for a heavy rail system.

Following a successful referendum, approval was given on 26th April 1946 to the initial Yonge Street subway between Union Station and Eglinton, and contracts were awarded beginning in August 1949. Much of the underground section of the route south of Bloor Street was built using highly intrusive cut and cover techniques, and the line was opened on 30th March 1954. At this stage the possibility still existed of a short east-west street car tunnel below Queen Street, and an interchange station was built at its intersection with the Yonge Street line. Expansion of the city, and ever-growing congestion, eventually led to this project being supplanted by a second heavy rail subway, the University-Bloor-Danforth line opened in 1966. Included in all these lines, and those opened subsequently, were purpose-built interchanges with feeder street car and bus services within a fully integrated system. An interesting reminder

Seen just after its enlargement by the London County Council in 1931 Holborn tram station in the Kingsway Subway shows a slightly art-deco style. Disused since 1952 this station and the adjoining ramp and tunnels are still in existence.

Södra Bantorget station, in the tram tunnel in Stockholm opened in October 1933, also shows a period style in its decoration. Converted to metro operation in 1950 the station is still in use, renamed 'Medborgarplatsen'. Note the overhead line for tram use, converted later to third rail; and the low platform.

of the street railway origins of the present subway system is that the original lines adopted the peculiar Toronto track gauge of 1495mm, although apart from a demonstration of new trains before the original opening no through operation has ever taken place. However Toronto retains an extensive street railway system, mostly on-street but latterly including some short subways.

Less complex subways developed elsewhere in the United States, in the form of relatively short terminal tunnels leading suburban cars to the centre. A short terminal subway was built in **Los Angeles** in 1925, allowing unimpeded access for cars from the Hollywood area. With the rundown of the Pacific Electric it closed in 1955. A further example was the 3.2km **Rochester** (New York) subway, opened in 1927 following the route way of the abandoned Erie canal, with a 4km cut-and-cover tunnelled section below a new street. The line was used by suburban street cars. There was a similar application in **Newark** (New Jersey) along the route of the Morris Canal, opened in 1935-7, and including a tunnelled section (the City Subway). In **Washington DC** there were short subways at C Street and the Bureau of Engraving terminus, and a longer underpass including a station at Dupont Circle, opened as late as 1949.

A late, and unique, addition to the roll of American tram subways came in 1963, in **Fort Worth**, Texas. This 1.1km line connected a large peripheral parking area to the basement of Leonard's department store in the city centre, with the inner part of the route in subway. It was worked by heavily-rebuilt PCC trams acquired from Washington DC. Most unusually in modern times this was an entirely commercial operation, without external financing. What is significant, not least in terms of the contemporary proposals in Brussels, is that at such a relatively early date hard-headed business interests in a car-orientated city saw advantage in a dedicated light rail link as a means of securing the future of central area retailing. After changes of ownership the shopping mall sadly declined and the subway eventually closed in 2002, although it had for nearly forty years demonstrated the potential of high-quality public transport.

Perhaps the most astonishing of the world's tram subways is one system which never opened. Construction began in 1920 on an 11km segregated railway in **Cincinnati** (Ohio), partly following another abandoned canal. By 1928 much of the tunnelling, four intermediate stations, and surface construction was complete but financial problems brought work to an end and it was never resumed. The substantial works remain in limbo today.

Between the wars the tramway system in **Stockholm** was notably progressive, and for many years pursued a policy of building segregated suburban prolongations to railway standards. To provide traffic-free access to the central area a tramway tunnel was built between Slussen and Skanstull in 1933, linked to existing tramways to the north. In 1941 the decision was taken to build a metro system and the first section was opened in October 1950, including conversion of the existing tunnel to metro operation, an ominous decision in Brussels terms. Some of the suburban lines were also later incorporated into the Metro, but others remained tram worked as feeders.

When the **Sydney** Harbour bridge was completed in 1932 it included, in addition to six road traffic lanes, four railway tracks of which those on the eastern side were used by three tram services from North Sydney, extended from their former ferry terminals. Leaving the bridge they descended into a short terminal tunnel, built for possible future use by main-line trains, leading to Wynyard railway station which they shared with the new 'City Circle' underground railway, and which bordered the main business district. There was an island platform with reversing sidings beyond, and the tunnel was block-signalled. With withdrawal of trams in 1958 the bridge tracks were converted into two additional road traffic lanes, and the tunnels leading to Wynyard became a car park and rifle range.

In **Oslo** two suburban feeder lines with light rail characteristics had fed the city tramway system, and in 1928 these were extended through a cut and cover tunnel to a new central terminus at the National Theatre.

A short terminal tunnel was built in central **Marseille** in 1893, initially steam worked. It remained in operation with trams until reconstruction and incorporation into the modern undertaking in 2007. It has been argued that the 3.7km *Földalatti* electric railway in **Budapest**, which opened in May 1896, was a pioneering tram subway and it indeed displayed some characteristics of later tramway applications: although largely underground it had surface operation on private right of way and its stations were located just below the surface, reached by stairways from opposite pavements. It was worked by single cars with overhead line power supply, and its tunnel was of restricted dimensions. However, it was self-contained and offered no running connections with surface lines. The objective of the line was to extend public transport through an area where surface running would be regarded as over-intrusive.

Stockholm was also notable for lengthy suburban tram extensions in what would become known as 'light rail transit' style. Most were later converted to Metro operation.

Köln's surface station on segregated track at Severinstrasse lies beyond the city centre subways and near the huge post-war Rhine bridge. This section was opened in 1970.

A quite different situation gave rise to a different sort of tram tunnel with little relevance to our story. Several of the world's tramways passed through tunnels occasioned by topographic features, to allow access to areas otherwise cut off by hills. The most striking current examples are in San Francisco, where the Twin Peaks Tunnel (3.65km, 1918) and the Sunset Tunnel (1.29km, 1928) expanded the street car network. The former included two intermediate underground stations. There were other examples on the Vicinal and in, for instance, Naples.

2.2 Germany's post-war plans

Given the extent to which **Germany** embraced the tram subway after the War —it has almost as many installations as in the rest of the world combined —it comes as a surprise that there were no significant examples before the 'sixties. Short underpasses in Berlin, Nürnberg, Duisburg, and Hamburg were all that were to be found: London and Stockholm had the most significant European installations before 1950. The concept of the 'second level' (*Die Zweiter Ebene*) for public transport was a German post-war formulation, and by the 1960s Federal Government funding policies in the West, together with the scale and scope of post-war reconstruction amidst the 'economic miracle', were producing a steady flow of U-tramway plans. One of the earliest was that in Frankfurt-am-Main, started in 1963, with Hannover beginning work two years later.

Whilst Belgian sources have tended to claim as their own the origins of the 'pre-metro' concept —of initial tram tunnels intended for eventual conversion—this is really true only of the terminology, although lines 1 and 2 in Brussels are the only examples (so far) of actual conversion taking place on that scale. To show that the theory was becoming widespread by the mid-1960s we need to glance at the following table. Conveniently the former (West) German Public Transport Alliance (*Verband öffentlicher Verkehrsbetriebe', 'VöV'*) published in 1965 a summary of tram tunnel and U-Bahn construction as it was then intended, covering most of the cities in the Federal Republic with populations over 300,000 (Wuppertal and Mannheim did not plan subways, and Hamburg and West Berlin emphasized extension of existing underground railways). It is important to note that, at that time, six of the twelve cities planned eventual conversion of tram subway operation to U-Bahn, although in the full sense this has not in fact occurred. We shall look more closely at some examples in the next section.

Tramways and Tram Subways in West German cities in 1965

	Route length km	Tunnel planned km	Conversion Planned?
München	122	0	-
Bochum-Gelsenkirchen	94	1	No
Bremen	59	4.6	No
Duisburg	80	4.6	No
Dortmund	97	15.0	Yes
Dusseldorf	160	15.0	Yes
Essen	98	5.7	Yes
Frankfurt/Main	135	32.0	Yes
Hannover	90	7.6	Yes
Köln	119	14.0	No
Nürnberg	68	11.5	No
Stuttgart	138	35.4	No

(Subways were eventually also built in Bielefeld, Bonn, Ludwigshafen, and Mulheim).
Source: VöV/Modern Tramway.

3. 'The Second Level' Tramway planning in Europe after 1945

Unexpectedly perhaps, a plausible timeline leads from Tremont Street, Boston (1897); through Kingsway (1902); and Slussen-Skanstull, Stockholm (1933) to Messrs Vane Morland in Leeds and Fitzpayne in Glasgow. They both wrote reports advocating tram subway development, in 1944 and 1948 respectively, and we know that Fitzpayne's circulated on the continent, including perhaps in Belgium. This may well have been the origin of the pre-metro.

The very model of a modern tramway: a Köln three-section articulated car crossing Barbarossaplatz in 1983, with segregated track and tram subways along its route.

3.1. Britain Leads the Way (but in vain)

Map 3. Eric Fitzpayne's post-war Glasgow plan showed the way ahead, ignored at home. Upgraded street tramways would have been linked to subways as well as to private tracks on ex-railway rights of way.

Given the subsequent history it is sad to find that two of the earliest detailed formulations of tram subways linked to high quality suburban feeders were set out in Leeds and Glasgow at the end of the Second World War. First was **Leeds**, whose Transport Manager William Vane Morland had visited Boston before 1939, and reported back on tram subways and PCC tramcars there. He had also visited the new subway installation in Stockholm. In July 1944 the Leeds Transport Committee made recommendations to the City's Postwar Reconstruction Committee advising a network of segregated-track suburban tramways linked by intersecting tunnels below the city centre. Leeds had already progressed some way towards achieving the first, with 'express' tramways extending east and south of the city, and further construction planned, some of which was completed in 1949. Detailed work was undertaken by the City Engineer, including trial borings and site evaluation for two or more tunnels extending east-west and north-south, with a two-level interchange station below City Square. Three new prototype single-deck tramcars were also acquired, intended to haul trailers (two are preserved). All these plans came to a summary end in 1953 with a snap decision to abandon the tramways, completed in 1959.

In 1948 **Glasgow**'s forward-looking Transport Manager, Eric Fitzpayne, produced what was surely one of the most prescient but neglected of all British urban transport studies. The plan was wide-ranging but for our purposes two recommendations are crucial, though neither took effect: he argued that existing railway rights of way, priceless in their penetration of urban areas, merited fuller exploitation and could better be used by conversion into electrified light railways using tram-type vehicles linked in part to existing tramways operating on segregated tracks; and extension of these light railways should take place both through city-centre subways and in the median strips of improved radial highways. (See map 3) Both these prescriptions, innovative in their day, have in fact since been carried into effect in various part of the world. In particular the modern pioneer of conversion of railway routes to light rapid transit use was Boston, with its Riverside Line in 1959, with a modest example in Brussels in the reuse of the former Tervuren railway, mentioned above, for an extension to route 39. Wider reorganisation of public transport around Glasgow (including the Glasgow Electric 'blue trains'), and adverse economic conditions, brought Fitzpayne's inspired ideas to an end and by the mid-1950s tramway abandonment had been decided, and all had closed by 1962.

However a knowledgeable visitor later noted in the head office of the Köln tramway undertaking a heavily annotated copy of Fitzpayne's Report suggesting that it had influence beyond its native land, and its full coverage in the technical press (including in the influential journal *Modern Transport*), suggests that it may well have been read in Belgium too.

3.2 Case Studies in Continental Europe

Let us now look at four European applications as case studies of the differing ways in which underground tramways have developed. They all were, like Brussels, 'legacy' tramways, that is to say, they had developed gradually over more than a century, and differed fundamentally from major 'new' tramways,

In the centre of Hannover in September 1983 conventional trams shared tracks with hybrid vehicles with folding steps which made best use of new subways.

such those in Nantes or Manchester, which had made a completely fresh start unaffected by previous constraints and standards.

Initially, despite abandonment of what would have been pioneering plans in Leeds and Glasgow, the conversion of the Stockholm tram tunnel to Metro, and the closure of the Kingsway Subway in London, post-war Europe appeared to be favourably inclined towards tramway modernisation. In this section we look at some key examples of tram subways, but note also the growing apparent threats from U-Bahn technology. Hamburg, a long-term local heavy railway operator, abandoned its tramways altogether, as did København (but without a replacement U-Bahn initially). In München, Wien, Nürnberg, and West Berlin underground railway construction seemed to be overtaking tramways, although in all but the last this was not completed.

3.21 Hannover

We think that it was in Hannover as early as 1949, not long after the formation of the independent Federal Republic, that the post-war concept of a system of underground tramways originated, in plans launched by Dr Philip Kremer, Technical Director of the transport undertaking. They were further refined by Professor Wehner of Berlin in a consultants' report in 1965. Investment on a large scale was justified on grounds similar to those set out later in Belgium: car ownership was swelling rapidly as post-war restrictions eased, and neither road nor parking space would be sufficient to cater for it

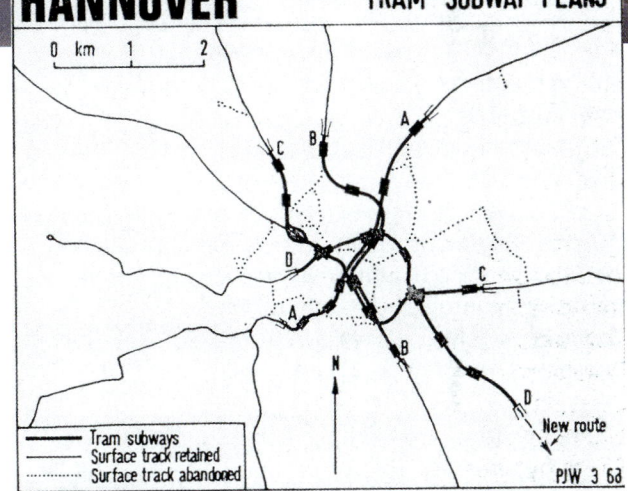

Map 4. The 1968 proposals in Hannover were partly realised over the next forty years.

without unacceptable destruction of the fabric of the city centre. Eventually a total of 36 route/km of tunnel was planned, and since 1965 much of this has been delivered. With Stuttgart (below) Hannover has most completely achieved the post-war ambition for transformed local transport.

The Hannover 'Stadtbahn' (as it is marketed) resembles those in Köln and Stuttgart, as it combines (in what we will see to be pre-metro style) subway with tramway aspects. Most of the central sections are in tunnel and elsewhere they are largely highway-based, although over three-quarters of the system is on segregated right-of-way. There are effectively three cross-city routes, all crossing at the city-centre station at Kröpcke. Map 4 shows the original plans, much altered later.

The first subway section (termed 'Route A') was opened in 1975 taking south-west/north-east services between Waterloo and Hauptbahnhof stations. Route

B is a north/south-east line, initially opened in 1979 between Werderstrasse and Kröpcke, and extended south in 1981 to Schlägerstrasse, and in 1982 on to Altenbekener Damm. Route C (north-west/south east) entered service between Kröpcke and Steintor in 1984 and continued west as far as Königsworther Platz in 1985.

The eastern branch of Route C, from Aegidientorplatz to Braunschweiger Platz and further on the surface, opened in 1989.The northern tunnelled branch from Steintor to Kopernikusstrasse finally followed in 1993.

A fourth east-west cross city line was envisaged (route D) but -as it had done in Köln (and Brussels)- surface tram operation continued in part of the city centre . Although the long-established street-running track to the Aegidientorplatz was closed after operation on 24th May 2017 this was not the end of street running. Construction is proceeding on new tracks leading to a terminus at the central bus station behind the Hauptbahnhof. These are expected to open late in 2017 and are intended to provide better interchange with the sub-surface lines.

There are short suburban tunnelled sections at Mühlenberger Markt, on route A in the southwest (lines 3 and 7 - extended to Wettbergen in 1998), and on the new southern section of Route D at Brabeckstrasse,

The Stadtbahn network now totals 123 route/km of which 19km are in tunnel (with 19 underground stations). Transfer between lines is readily available at Kröpcke, Aegidientorplatz and Hauptbahnhof , and there are many efficient inter-modal interchanges providing an integrated system.

Hannover was also, with Frankfurt/Main, a pioneer of hybrid cars with folding steps.

3.22 Köln

The important Rhineland city of Köln suffered as much as any from wartime destruction and in the course of its rebuilding provision was made in a new street pattern for segregated highway-based tramways.

As in Brussels and other cities conventional underground railways had once been considered in Köln, but the expected traffic flows never justified the huge capital cost and nothing happened before the Second World War. In conjunction with the comprehensive redevelopment plans which followed a survey of city traffic requirements in 1956, urban rail development was endorsed but without specifying the form. Further studies by Professor Dr Lambert beginning in 1960 produced firm plans which were approved by the City Council in 1961. These envisaged a coherent system of central tramway subways (rather than a series of underpasses on surface tracks). As in Brussels there was passive provision for ultimate conversion to U-Bahn operation, with temporarily-lowered platforms and clearances for 2.65m-wide stock (although this is not now considered a likely prospect). Diversion of utilities began in 1962 and pile-driving in September 1963. The system would resemble an 'H' on its side, with east-west lines running from the river west towards Bocklemund in the north and to Barbarossaplatz in the south, with a north south line through Neumarkt linking them together. A second long north-south line would run closer to the west bank. Several short branches led to suburban ramps, and the whole system was joined to the east bank by segregated tracks across two Rhine bridges. The result was a comprehensive system within the old centre, with longer tunnelled

A modern subway portal, in this case near Zoo-Flora station in Köln, opened in 1974.

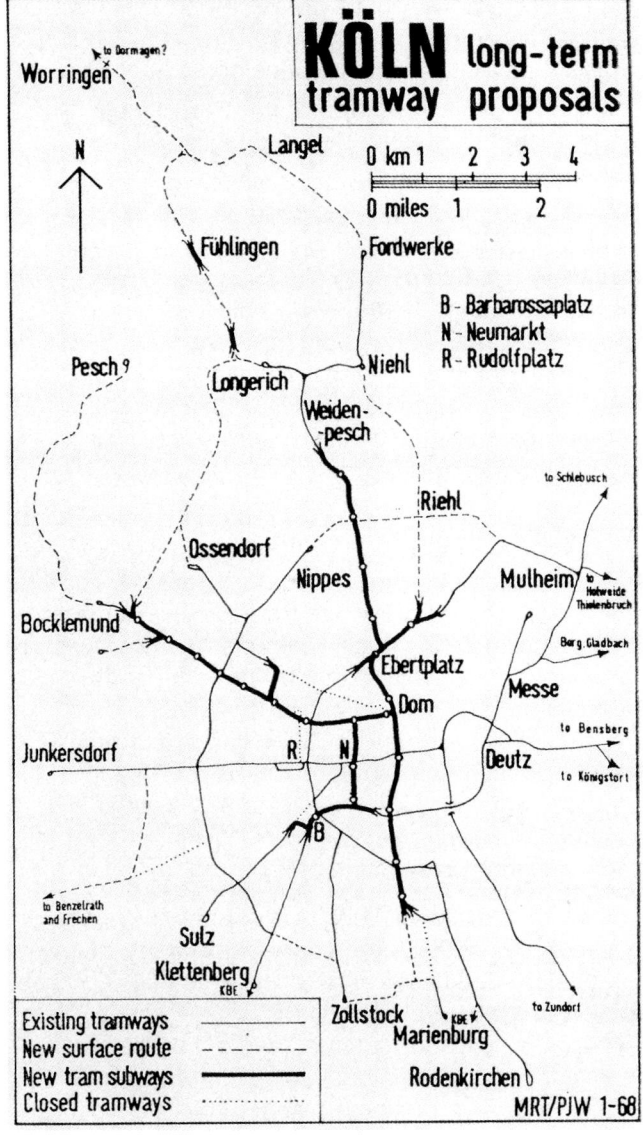

Map 5. Köln proposed a comprehensive net of tram subways, but retaining some surface lines where street conditions allowed, and linked to segregated tracks in the suburbs. The actual network extended further.

extensions west and north. The first part of the 7.5 km initial system opened on 11th October 1968, and was extended between 1970 and 1974. Further construction took place between 1983 (in Deutz) and 2015. It continues today, after an engineering catastrophe which delayed work on a new north-south link. Map 5 shows the original intentions, which with modifications and extensions have been largely fulfilled.

Where Köln differed from other systems, including Brussels and Stuttgart, was that a significant presence of surface tramways remained within the central area, and to an extent still does today. Importantly much suburban track has been transferred to reservation, and elsewhere some sections have been built on elevated structures rather than in tunnels. Shared space with road traffic is now unusual, and this impacts favourably on reliability and average schedule speed.

Subways also extend to and through the east-bank suburb of Deutz. Over what was formerly the Köln-Bonner Eisenbahnen interurban line, the system reaches south to link with the Bonn Stadtbahn which includes subway sections.

The result after nearly fifty years of systematic construction is a very high quality transport system, largely protected from surface traffic congestion. The system now totals about 195 route/km, with about 40 underground stations.

3.23 Wien

We move south now to the Austrian capital, which was (and is) one of the great 'tram cities' of the world, with the fourth-largest current network. Like Brussels the city's layout in part reflects the previous existence of city walls and their replacement by a broad circumferential highway. Unlike Brussels however, and despite the inauguration of an extensive U-Bahn system after 1978, the conventional tramway has retained a strong city centre presence and has continued to expand despite some replacement by underground routes.

There had been proposals for sub-surface tramways in the city as early as 1902, following on from the success of the 1896 line in the Empire's 'twin' capital, Budapest. Another such initiative in 1909 came to nothing before the Austrian catastrophe in the Great War. Another parallel with Brussels, not unlike the 1957 Midi tunnels, was the opening of short tramway underpasses and stations at Sudtiroler Platz in 1959 and at Schottentor in 1961, which provided a first glimpse of what a 'second level' might provide.

The wider tram undertaking might have developed into a classic 'pre-metro' system after the Second World War and the end of allied occupation in 1955. In 1960 plans were published for such a network, crossing below the old city and supplementing the existing 'Stadtbahn' lines around its periphery (see map 6). There would have been three crossing and interlinked tram tunnels with several branches, handling many existing services and connecting to surface lines at twelve ramps. In November 1963 the first work began on a 1.9km tunnel between Alser Strasse and Karlspaltz, beneath the succession of streets known as 'Lastenstrasse', a peripheral route to the west of the centre; there were four intermediate stations. Tunnel dimensions allowed for eventual conversion to U-Bahn operation, but the low-level and short platforms at stations would require modification. This tunnel opened to trams on 8th October 1966. It handled three different services and thus offered direct connections to and from a range of suburban destinations. The tunnels were permanently lit but simple block signals were installed at 100m intervals.

Construction of a second and more elaborate section of tram subway began in 1964. This 2.3km line lay to the south of the centre, along the southern portion of the 'Gürtel', an inner ring road. It included six stations, and in addition to the main east-west trunk had three branching sections and five ramps.

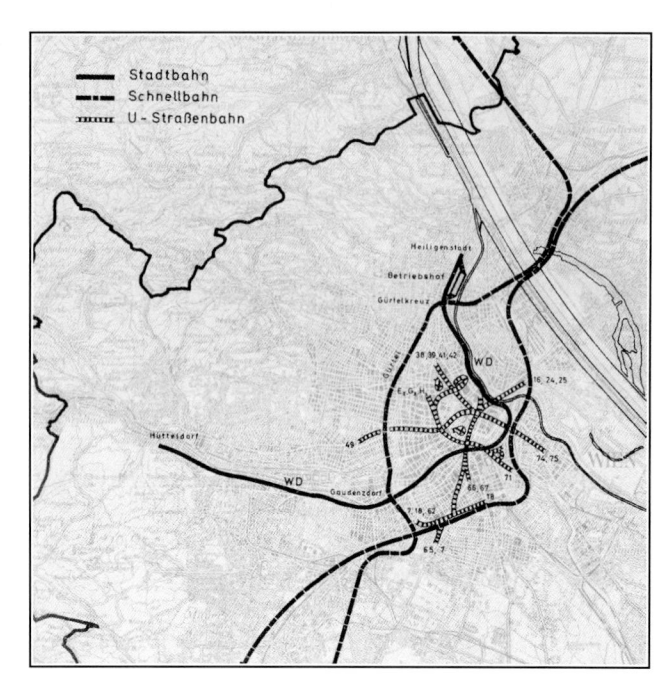

Map 6. In 1960 Wien announced plans for a dense net of city centre tram subways linked to many surface lines. Instead a full U-Bahn plan was adopted six years later, though many tramways survive, even in the centre.

Opened on 11th January 1969 the line again had low platforms only 70m long. In the original concept the northerly Laurenzgasse branch would have connected at its northern end to the main subway net under the old city. Even before the completion of the Gürtel tram tunnels a decision had been taken in 1968, after some controversy, to construct a U-Bahn system with three lines. The Lastenstrasse tunnel was to be incorporated into an early stage of this network . The new system would also incorporate some of the long-established Stadtbahn partly elevated system, opened with steam traction in 1898. Opposition, as in Brussels, centred on the ending of direct one-seat service to and from the suburbs, and the need for more passengers to change modes.

Construction of the U-Bahn system began on 3rd November 1969, with the first section of line U1 opening on 25th February 1978, between Karlsplatz and Reumannplatz. The 'basic network' was completed on 3rd September 1982, and with more recent extensions the system is now made up of five lines with 83.1km of route, serving 109 stations, following a further extension in September 2017. As in Stockholm and München some U-Bahn extensions have replaced surface 'rapid tram' lines, but there has been similar public criticism to that in Belgium regarding increases in walking time and interchange time when parallel street-based services are withdrawn.

Upgrading of the Lastenstrasse tram tunnel continued during tram operation, although adaptation of the stations was problematic. The tunnel was extended at both ends, to an overall length of 3.6km and with two additional stations, and opening to U-Bahn trains took place in August 1980. Meanwhile about 30 route km of tramway have closed since 1971, and the system now amounts to about 180 km with about 30 different services. There has been no additional tram tunnel construction, but there are no proposals to convert the Gürtel tunnel to U-Bahn use.

Although Wien's tram system is still extensive, and is being re-equipped with new low-floor rolling stock as well as extended into new areas, the U-Bahn now dominates much cross-city and suburban travel, and as in Brussels the tramway is partly retained in a feeder capacity. Arguably the integrated Wien network is amongst the most successful in Europe in attracting and retaining traffic: 74 per cent of commuting is by public transport, compared with about 50 per cent in Brussels.

3.24 Stuttgart

is our last example, a hilly and prosperous city and unlike the other two German tramways we consider it almost wholly rebuilt its traditional tramway over forty years, to the extent that it rather misleadingly now claims to have a 'U-Bahn'. After restoration of the city following severe wartime damage Stuttgart had an extensive and comprehensive metre-gauge tramway system, re-equipped in 1959-64 with a fleet of about 350 attractive GT-4 type articulated cars capable of operation in multiple.

The present 'notional' U-Bahn is not how the plans began. In April 1961, after considerations which would have been familiar in Brussels, the City Council approved a plan prepared by the undertaking's Director Professor Bockemuhl (one of Europe's leading tramway engineers), for a complex system of tram subways, originally planned to replicate much of the surface tram network in the central area. Subsequently revised and simplified the plans eventually envisaged 9.5km of double-track tunnels, 3.3km of single line tunnels, 1.17km of double-level tunnels, and 16 underground stations, some of considerable complexity. Work began in 1965, the excavations made more elaborate by the simultaneous construction of road tunnels and storm water drains. Allied to improvements to connecting surface tramways, where possible introducing segregation from road traffic, what was planned resembled in principle the later formulation of the Brussels pre-metro. Dimensions here too would have allowed for eventual conversion to standard gauge operation, although this was not at first proposed.

Changes in senior personnel and of political influences led to changes long before the 1960s plans were complete. In March 1965 the Council revisited the plans in terms of a partial conversion to full-scale U-Bahn operation. This would be undertaken (as in other major West German cities) alongside a related network of main-line railway routes branded as an 'S-Bahn'. Together a coarser net of new U-Bahn lines, together with an underground S-Bahn net with several branches, would have allowed complete replacement of tramways, and especially of surface-running routes. Whilst this master plan went no further its shadow lingered on, and the very high cost of subway construction led in 1969 to some scaling-back, for example by eliminating some of the complex junctions. Construction in fact proceeded over more than three decades, with the first subways opening in five stages between 1966 and 1983, initially with metre-gauge operation. As in Brussels temporary ramps enabled operation to continue section by section. In 1976 the formal change to a 'Stadtbahn' was agreed, together with conversion to standard gauge after a long and complex period of mixed-gauge operation (the case for this was perhaps questionable, given successful metre-gauge operations in Zurich and elsewhere). The first of an initial batch of standard-gauge cars was delivered in 1978.The first complete route conversion followed on 19th April 1986, when the 'stadtbahn'-type cars replaced metre-gauge trams between Fellbach in the north-east of the city and Vaihingen in the south-west. Initially the whole route was mixed gauge. A route between Vaihingen and Mühlhausen followed in July.

Completion of the standard-gauge 'U-Bahn' network was expected by about 1995, with some metre-gauge routes expected to remain as tramways even after that. The cost of conversion, together with the aspiration to end street-running (not wholly achieved even now), led to significant reduction in the coverage of the network, and rather than a branching network the system has developed partly into a web of individual routes in U-Bahn style. As in Brussels subway stations included lowered platforms during the tram-operation phase.

Metre-gauge rolling stock, still relatively new, was replaced by larger and heavier cars, each 2.65m wide, each unit accommodating 234 with 110 seated. The hilly terrain of the city demanded powerful traction equipment and each unit had four 222 kW motors. This fleet of elegant new cars was delivered between 1985 and 2003.

Plans to retain some metre-gauge routes failed and by 2002 conversion to standard gauge marketed as a 'U-Bahn' was virtually complete. Segregated surface track has largely replaced street running with high-level platforms where possible. Although tramway advocates at times decried the manner in which the network evolved, and especially the concentration of traffic on fewer corridors, the end result was not notably dissimilar from that foreseen in Brussels in the 1960s, and is arguably the apotheosis of the pre-metro concept in its balance of tunnel and street-based tracks. Stuttgart now has an enviable 116 -km system of high capacity rail routes, largely insulated from traffic congestion and offering high-quality service at impressive schedule speed. The network is less comprehensive than it could perhaps have been, and the cost of achievement would have been inconceivable other than in a very rich country, but the end result is a fine vindication of the 1960s principles which we shall explore in the rest of this book.

The ramp at Schlosstrasse in Stuttgart had just been completed in September 1983, and metre gauge cars were still in use although mixed gauge track had been installed.

4. The Belgian and Brussels Context from 1945
'Le Carrefour de l'Occident'

We now turn to specific opportunities and plans in Brussels itself where the pre-metro originated, a metropolis in transition in the post-war years, and the scene of several planning innovations. We first review the state of the city after 1945.

In the 1950s the old centre of Brussels retained its historic charm, although tram transit was slow and difficult. [Courtesy Patrick Sellar]

4.1

Brussels in 1945 was still a relatively modest capital of a small nation, but one which already punched above its weight in international affairs. Partly this was a result of Belgium's location at a crucial crossroads in Western Europe, a position which had been fatal in 1914 and 1940. In the railway age Belgium was sometimes called 'the turntable of Europe' because of its nodal position in the Continent's railways. Then almost from its independence in 1830 Belgium had become an industrial and commercial powerhouse, the first continental state to embrace large-scale industrialisation to the extent that by 1900 it was second only to the United Kingdom in terms of national wealth per capita. It was the 'Asian Tiger' of its day. Finally, riches had flowed into the country, and into Brussels and Antwerpen particularly, from Belgium's controversial African possessions which contained major sources of minerals and forest products. It was industrialisation and Africa which helped to build the city's enviable infrastructure.

But in 1945 these achievements were somewhat dimmed. The Great Depression had hit the country particularly badly – following the Wall Street Crash in 1929 Belgian unemployment had risen by 1932 to no less than 23.5% of the insured population, and state revenues had fallen by a third —and defence expenditure had grown massively in the 1930s in preparation (vainly) for the expected war. The occupation from 1940 to 1945 was not only vicious and destructive but also leeched out Belgian assets and products for the German war effort. Although Brussels, unlike British and German cities, suffered relatively lightly from aerial bombardment it became run down, shabby, and conflicted. The aftermath of victory in May 1945 included not only a settling of scores with those believed to have been collaborators, a process which continued for several years, but also bitter political and social division over the role of King Leopold III during the War, leading to his abdication in 1951 amidst significant disorder. Putting this behind the nation, and embarking on a new and optimistic course under a new King, was a major political imperative in the years that followed, and some of what we will consider was part of that process. In 1945 Brussels had national interests to pursue.

Ten years later it was had begun its transformation into a hub of international institutions, beginning with the establishment of the customs union of 'Benelux' in 1944, and the Treaty of Brussels in 1948 which set up the Western European Union as a defence pact embracing the United Kingdom, France, Belgium, Luxembourg, and the Netherlands. The European Iron and Steel Community followed in 1951, and the European Economic Community (later 'European Union') in 1957. The North Atlantic Treaty Organisation moved to Belgium from France in 1967. All these institutions built a major presence. They and their associated bodies have made 'Brussels' a personification of supranational organisations, to be reviled or respected according to convictions, but there is no doubting that they, and their spin-offs in terms of supporting service activity, have profoundly altered the appearance and culture of the city and incidentally its transport needs.

The period after 1954 was also a period of relative political stability in Belgium, which fostered longer-term planning. With some foresight Omer Vanaudenhove, then Minister of Public Works, spoke as early as 1956 of Brussels becoming '*le Carrefour de l'Occident*' (the crossroads of the West), a lofty aspiration which underlay much of what followed.

Unlike most metropolitan entities Brussels had until relatively recently no over-arching city government. It was itself a federal structure, embracing nineteen different communes or boroughs, which retain important local responsibilities, including primary education but not public transport. The 'City of Brussels' ('*Ville de Bruxelles*', the national capital) is itself only one of the constituent parts, rather like the 'City of London': it embraces the centre of the urban area, a substantial area to the north-west around Laeken, and (for historical reasons) a long southwards antenna along Avenue Louise towards the Bois de la Cambre. Since full regionalisation in 1988-9 the 'Brussels Capital Region' has become an enclave of 162 square kilometres within the Flemish Region of Belgium, with a large measure of devolved government, including overall responsibility for local transport. For topographical and historical reasons the Region is slightly lopsided, with the central area (the 'Pentagon') located towards its western edge, and the extent of the built-up area greater to the east and south. As seen from the maps, this imbalance is reflected in the shape of the tramway and Metro system today.

Although the first and second Kings of the Belgians had whole-heartedly striven to establish a distinguished and elegant national capital from what had been a large Flemish city (much the same size then as Gent and Antwerpen), construction had stalled after 1914, and apart from impressive housing developments around the periphery little had been realised later in terms of additions to street resources.

4.2

There were three interlocking factors which led directly to an upsurge in transport infrastructure beginning in the 1950s. These were:

• The celebratory national dynamism seeking to expunge the humiliation and loss of 1940-44, and with it a determination to replace substandard housing and to resource new industrial and commercial activities to banish the spectre of unemployment which had blighted Europe (and fostered political extremism).

• The long-delayed completion of the 'Junction' railway link between the Midi and Nord stations.

• The decision to hold a major international exhibition on the Heysel Plateau in 1958.

The post-war concept of the Cité Administratif accompanied the Junction Railway works, and was partly achieved. Developments like this swelled the demand for public transport on this axis.

A futuristic dream in the 1960s later became reality: this is the original concept of the 'city of towers' of the Espace Nord, Brussels' 'mini Manhattan'.

A reminder both of the massive destruction caused by the building of the Junction Railway, and of the dereliction left for decades whilst the work was suspended. We are looking in 1950 towards Gare du Nord not far from the present Congrés station. A tram can just be seen in Rue Pacheco, a line abandoned in 1953. This is now the site of a broad new boulevard.

In the national context it is worth noting that local transport was a small part of a vast infrastructure programme. Beginning in the late 1940s Belgium created a comprehensive motorway network, electrified main line railways, improved and enlarged inland waterways, and modernised and expanded docks.

Although Belgian heavy industry would falter in the years to come, as everywhere in Europe, the nation was quick to develop other export opportunities in fields such as chemicals, electrical equipment, and automobiles which contributed to a consistently high level of economic growth for over twenty years, providing the means to pursue a consistent programme of infrastructure investment. The shocks which followed the oil crisis of 1973-4 brought this dynamic process to an end and profoundly affected the realisation of the plans drawn up in the 1960s, as we shall see. But let us first consider the effects of the three factors mentioned above.

4.21 New government and commercial quarters

With tolerant planning regulations and building controls for the first fifteen years after the War, ample capital, and suppressed opportunity for investment, Brussels was for over twenty years a fertile ground for building speculation, and its face was permanently changed by massive new constructions, not all of the highest quality. Three examples will indicate the scope of activity and also the demands it placed on public transport. The completion of the Junction railway (see below) offered opportunities for a swathe of new building along its route, notably in the northern section where the tunnels were covered by a broad new boulevard. Amongst new buildings here was a vast centre for public administration, a new 'Whitehall' for Belgium, '*La Cité Adminstratif de L'Etat*'. This was intended to gather together from scattered and unsuitable buildings the growing activities of the national government. First considered in the 1930s a decision to proceed was taken in 1948 and architects were appointed in 1955. It was expected that upwards of 7,500 civil servants would be relocated, especially in the areas of public finance, public health, social security, and communications. The work would take 25 years to complete. The new 'Congrès' station on the Junction railway was located close to the development, but the scale of activity and of ancillary enterprises would create new and heavy traffic flows on the tramways too.

The second example, by contrast, was entirely commercial in origin. By the middle of the 'Golden Sixties' available capital and seemingly endless demand for office space spurred comprehensive redevelopment in many cities, and in Belgium very low corporate taxes encouraged foreign relocation and injection of capital. One resulting development was intended to adjoin the new Nord station on a site of over fifty hectares: although required for practical reasons the relocation of the station had the effect

A striking view of Gare du Nord from the south shows how the rail tracks rise from their tunnel (right) and curve into the new station located well to the north of the original. The 1950s clock tower is dwarfed by new high-rise buildings.

of opening up a new commercial quarter around it. Over 10,000 people would be moved from their homes and new tower blocks and elevated walkways would provide commercial space, and also a new administrative headquarters for the City of Brussels. A new '*Communications Centre Nord*' extended the Nord station, including bus terminals and a pre-metro station beneath. The plans were substantially scaled back after the financial crisis of 1974 and were not resumed, in altered form, until the 1990s, but the travel demands of the whole quarter have profoundly influenced the development of traffic along the north-south tunnel.

Finally the growth of the European institutions following the establishment, and later enlargement, of the European Economic Communities provoked massive development to the east of the city along the Rue de la Loi, notably the Berlaymont Building (1967-70) near the Schuman intersection. This, with later and larger buildings including the European Parliament, housed over 20,000 workers: this transformed travel demands along the west-east corridor and affected transport infrastructure. These developments were supported by commercial developments to house activities related to the Parliament and the Commission.

4.22 Completion of the 'Junction' Railway

Joining the separate Midi and Nord terminal stations in Brussels had been an ambition for over a century and was the subject of recurrent national controversy. We have become used today to the obliteration of parts of our towns and cities by vast arterial roads and urban motorways but for its day the Junction was a unique and awesome project.

The construction of the link must have been one of the longest-lasting such projects in modern European history. Conceived in the nineteenth century and approved in 1903 work began in 1911, but after acquisition of property along the route and some construction, including part of the viaduct just north of Gare du Midi, the invasion of Belgium in August 1914 brought the work to a summary end. Recommenced in 1919, it was later cancelled again following post-war economic crises together with disagreement between different national interests. Bizarrely for ten years the partly-finished viaducts, and stretches of cleared land, stood fallow, forming an ugly scar across the old city (see illustration on page 26). Partly in order to provide employment during the next economic crisis a new government agency was formed in 1935, the *Office National pour l'Achèvement de la Jonction Nord-Midi* (ONJ), and work began again, only to cease for the third time following the invasion in 1940. Perhaps surprisingly the Germans (confident of imminent victory) permitted building to recommence in 1941, continuing for another two years before general shortages of material and labour brought it largely to an end in 1943. Work to complete the Junction finally began again at the end of the War.

The six-track line, 35m wide and 3.6km long, is built on viaduct and in cut-and-cover tunnel, and its construction cut a massive path along the eastern sector of the city's historic centre, curving to avoid some of the more important historic structures (such as the Collegiate Church of St-Michel & Ste-Gudule, now the Cathedral) but destroying vast areas of old buildings in a way inconceivable today. What would now be a precious site, the former girls' boarding school attended by Charlotte and Emily Brontë in 1842 and where Charlotte later taught, was amongst the lost treasures. Part of the project was the construction of new roads and business premises above the tunnel, innovations which strengthened the economic case for the line.

There were three new intermediate stations, at Chapelle and Congrès (for local traffic only), and a new underground Central Station well located close to the old centre. Each of the former terminal stations was relocated. The Midi station was rebuilt slightly to the south of its original site and elevated from its original street level by some 20m to match the Junction viaduct completed decades earlier and idle ever since. As the station was situated close to the Petite Ceinture, with heavily-trafficked streets on three sides, the relocation of tramways was a major part of the reconstruction project.

The new Nord Station, not completed until 1956, was relocated several hundred metres north of its previous convenient site facing the once-handsome Place Rogier. The design of the station included a plaza above traffic level immediately adjoining the huge concourse on the west, and tracks were laid into an open-air tram and bus station at the main entrance. The location of this station, remote from the main east-west traffic flow along the Ceinture, has been a weakness ever since: its displacement was made necessary because the Junction tracks had to pass beneath the Ceinture on leaving their tunnel, and the acceptable gradient necessarily displaced the station northwards.

The railway itself was finally opened by H.M. King Baudouin in October 1952, although it was not to be wholly complete until 1954.

4.23 'Expo '58'

The Brussels 'Exposition Universelle' of 1958 was a major event in the city's post-war history, and left a lasting mark on the tramway system. The initial proposal was made by Mayor Joseph van de Meulebroeck in 1948, and a planning commission was appointed in 1951, with an estimated budget of 1,080 million Belgian Francs, partly financed from the colonial Lottery.

The site chosen –the Heysel plateau north-west of the city centre—was that already created for the delayed centennial exhibition of 1935, for which major new tramway infrastructure had been provided. 'Expo '58', the first major international event of its kind in Europe since the end of the war, took place in a spirit of great optimism in which Belgium, and Brussels, successfully sought to promote their style, modernity, and efficiency. The provision of high quality public transport was part of this mission, but also had the essential purpose –at a period when widespread private motoring was becoming established- of allowing mass access without producing impossible traffic problems. Work on the Expo site began in 1955.

Heysel was remote from main line railways so all national and long-distance passengers would have to transfer to trams or buses to complete their journeys. Both tramway agencies, STIB and SNCV, were involved. The plans had three elements. Firstly, tramway service was enhanced by the provision of new vehicles and by improved tramway

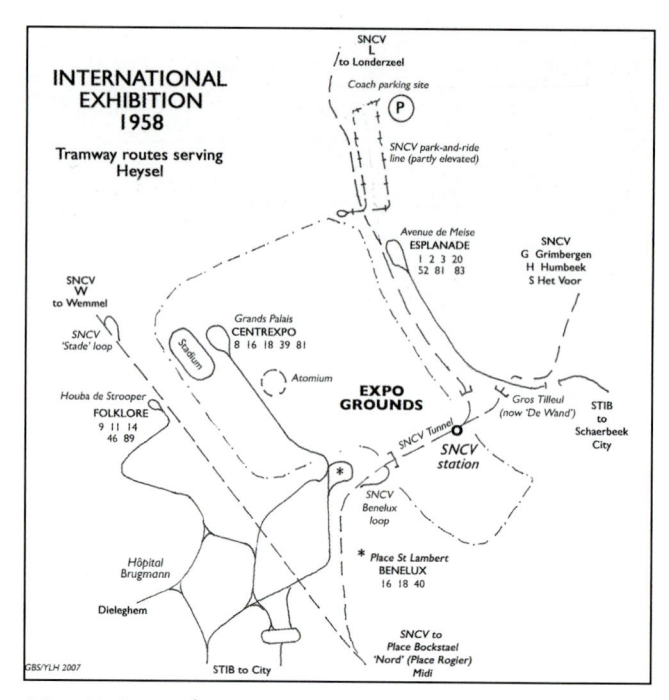

Map 7. Brussels Expo 1958: tramway arrangements around Heysel.

The Vicinal tunnel at Heysel, built in 1957, was a major work: this is the south ramp under construction, leading out of De Wand station. [Courtesy LRTA Collection/Online Transport Archive]

Not long after the tunnel opened an SNCV tram from Londerzeel descends the northerly ramp towards the city centre, and will soon stop at the Vicinal's underground station. Unlike the rest of the subway this section has not reopened. [Courtesy LRTA Collection/Online Transport Archive]

Side by side: trams of the two competing Brussels operators call at De Wand in September 1958, with a Vicinal 'S-type' car from Strombeek to Place Rogier on the right; and a motor-and-trailer STIB set on service 1 on the left, coming from the city and Bois de la Cambre. The station is still in use today.
[Courtesy Tony Percival]

facilities. Secondly, much work was undertaken to improve general traffic circulation around the city by the provision of wider roads, underpasses, and segregated highways (which contributed to freer-flowing tramway traffic). Thirdly, there was provision for tram-served 'park and ride' near the exhibition site. Map 7 shows the 1958 arrangements.

Improvement of the 1935 tramway facilities affected both SNCV and STIB, and the former's line along the south-east of the site was radically altered. The exhibition site had been enlarged by some 200 hectares to the south-east, extending into the Parc de Laeken adjoining the Royal Palace. This enlargement necessitated a substantial reordering of the metre-gauge lines in the Rue du Heysel, which had in effect formed the south-eastern boundary of the original grounds. After protracted use of specially-built deviations, SNCV services were diverted into a new 520m tunnel, used by service L on the branch to Londerzeel (which diverged inside the tunnel itself) from February 1957, and by the 'main line' Grimbergen, Humbeek, and Strombeek services from September 1957. In addition a new SNCV tram station, turning loop, and storage sidings were built close to the 'Benelux' gate at Place St Lambert. An attractive new SNCV building, with generous shelter, a small office, and a waiting room, was built close to the exhibition gates. Furthermore an underground tram station was built within the tunnel, at the divergence of the Londerzeel and Grimbergen lines and close to the 'Nations' gate and the huge US and Soviet pavilions. This was in fact the first tramway tunnel to be completed in Brussels, and although it

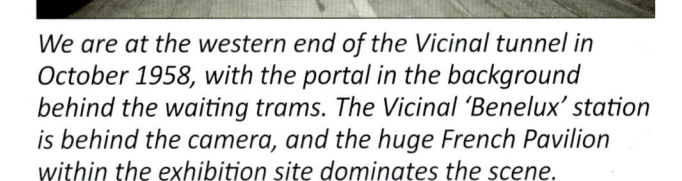

We are at the western end of the Vicinal tunnel in October 1958, with the portal in the background behind the waiting trams. The Vicinal 'Benelux' station is behind the camera, and the huge French Pavilion within the exhibition site dominates the scene.

A fine view in Blvd Leopold II west of Place Rogier in September 1958 shows how through road traffic on the northern sector of the Petite Ceinture was carried above the trams on a lengthy viaduct. Standard-gauge STIB tracks lie on the left (with a single motor inbound from Ganshoren on the 13 service), and a Vicinal train on metre gauge on the right, nearing its 'Nord' terminus on the way from Londerzeel. We shall see later how this section was replaced by tunnels in the 1980s. [Courtesy Tony Percival]

29

was abandoned when the last Vicinal services were withdrawn in 1978 it was later restored by the STIB and is in use today: the underground station was not reopened.

The STIB line to the north-east of the site formed part of a lengthy extension originally built in 1935 alongside Avenue des Croix du Feu. This was extended beyond the original 'Gare Astrid' terminus in 1957 to a new terminus adjoining the Exhibition's Esplanade gate, and an SNCV stop on the Londerzeel line was nearby. It was normally served by STIB routes 1 (from Bois de la Cambre at the end of Avenue Louise, south of the city) and 52 (from Drogenbos, in the extreme south, passing Nord station). Extra unnumbered services were provided for part of the season.

The Esplanade route was greatly improved by new underpasses at Gros Tilleul and Pont Van Praet opened in 1957. In addition the tracks the length of Avenue des Croix du Feu were transferred to side reservation, and a new low-level station was built at the location now called 'De Wand', incorporating the SNCV tracks serving Grimbergen and beyond. The STIB and SNCV tracks, which had formerly been on mixed gauge west of here, were now wholly separate. The reconstructed route from the Schaerbeek district to Esplanade became one of the most impressive on the system, almost entirely segregated from road traffic and making possible rapid and uninterrupted travel to and from Schaerbeek railway station as well as from the northern part of the city centre. It has since been further improved and forms part of 'quasi-metro' line 3.

Compared with 1935 far greater use by exhibition visitors of car and coach transport was expected, and large privately-operated car parks were built some distance from the north end of the site. The SNCV tendered to provide the internal park-and-ride service from the major peripheral coach park, situated alongside the Antwerpen motorway at the north-east extremity of the site. The company operated a short new uni-directional line, built by the Exhibition's parking contractor Tedesco, running mainly on a skeletal concrete viaduct alongside and across the Avenue de Meise, and terminating close to the Grands Palais exhibition entrance. This connected for operational purposes to the SNCV's adjacent Londerzeel line and was served by a dedicated service of motor-and-trailer sets, normally with three sets in use during the week and four on Sundays. The SNCV's contract had specified a maximum service of nine sets in service, carrying up to 8000 passengers an hour at a headway of two minutes, although generally traffic fell short of this perhaps because of the high parking fees payable (one of few subjects of public acrimony connected with the exhibition). The line was dismantled in 1958.

The Exhibition was the pretext for a near-orgy of road construction in many parts of the city, the impact of which permanently changed the character of several main thoroughfares. The most affected section was the inner ring or Petite Ceinture, where underpasses were constructed at major intersections on the eastern side of the ring, and a new elevated highway was built along 1.25km of the northern edge west of Place Rogier. Similar underpasses were built the length of the Avenue Louise. Necessitating the removal of thousands of mature trees, the effects of these alterations have even now not been wholly softened. Although no tramway tunnels were built here as part of this project, the circulation of trams was substantially improved by the removal of through road traffic, for example at Place Louise and west of Place Rogier.

Whilst much of the exhibition was dismantled after 1958 its site remains an important venue for sporting, entertainment, and business purposes, and the transport links created in 1935 and 1958 continue to support this.

4.24 Hospitals, Offices and Universities

A further fundamental change in the urban fabric, with important consequences for public transport, was apparent throughout the Western world from the 1950s. This was the expansion, and in some cases relocation, of public institutions such as hospitals, schools, universities, and further education institutions. Office functions also ceased to be solely a feature of city centres and some moved to the suburbs. In Brussels the progressive improvement of the line 5 pre-metro, and the extensions to the catering college 'CERIA' (1951), Erasme Hospital (1982), and Jette Hospital (2018), amongst others, reflect such trends.

4.3 The Midi Subways

We now need to look more closely at the Midi station area to determine the purpose and execution of the STIB's first tram subways there, deriving as they do directly from the 'Junction' project. They began neither as preparations for the 1958 exhibition nor as a first phase of a larger tunnel plan: the project was financed as part of the railway budget and planned well before consideration was given to any general tramway improvement.

The railway itself was finished in 1952 but at this stage the standard gauge tram tracks still crossed the busy Boulevard du Midi on the level into Boulevard Lemonnier and Boulevard Poincaré. When planning the resumption of work on the Junction this problem of conflict between trams and growing road traffic around the new station had been foreseen. The original intention, however, was not to sink the tramways underground, but to build several relatively small and short road underpasses, leaving the trams to circulate on the surface. One such short road underpass was in fact built at the inner end of Avenue Fonsny on the east flank of the station but has since been removed. By 1948 the present solution of a complex underground tramway junction with four branches had been adopted. Plan 8 shows the initial layout, part of which survives today (see also map 15).

In the 1940s the standard-gauge tramways passed the station on three sides, to the west in Avenue Fonsny, to the east in Rue de France, and in front in Place de la Constitution. There was a turning circle on the east side, and SNCV tracks also passed in front of the station. It is impossible now to visualise the ambience of the station as it was before the Junction works began in 1911. Place de la Constitution was a regularly-shaped expanse, bordered by unpretentious but agreeable buildings, many of them housing hotels and cafés. On the south side was the monumental front of the station itself, with its prominent classical portico surmounted by a massive sculptural composition. The centre of the square was occupied by lawns, gardens, and trees with a taxi-rank alongside. All this had been dominated by the unfinished Junction viaduct to the north, but what remained of the gracious layout was now to be swept away completely.

The construction work went through nine phases, beginning in June 1948, each requiring extensive relocation of tracks, mostly temporary. An important diversion in 1956, needed to allow excavation on the site of the ramp which would lead north into Boulevard Lemonnier, took trams away from the boulevard via Rue de Woeringen and into Boulevard du Midi. (A similar but longer diversion would be required when the north-south tunnel was under construction.)

For over seven years interchange here between train and tram was makeshift and ever-changing, and must have caused considerable trouble to both passengers and the STIB staff. Not until July 1954 was the new passage beneath the railway adjacent to the concourse ('Rue Couverte') brought into use and it was a further three years before the four-track layout and platforms there were ready.

The site of the Place was partly obliterated by the newly-completed viaduct into the station. The new station itself was moved some distance southwards and with a broader footprint. It was also raised some metres to match the new viaduct approach from the north. The entrances were located on the east and west sides, with the principal entrance to the east beneath an imposing tower. Both the Midi and Nord stations now had towers, though the former has been demolished and the latter is now dwarfed by the surrounding blocks: their intention was to advertise from afar the presence of the relocated stations. For a construction site extending over only some 750 metres the period of work on the tram tunnels will seem long, but the subterranean task was immense. Sewers, cables and gas mains were everywhere, and the foundations of older buildings were found. Pockets of water and gravel were unexpectedly discovered. Chemical consolidation of soil was needed close to other buildings. The tunnel was roofed with I-section steel beams reinforced

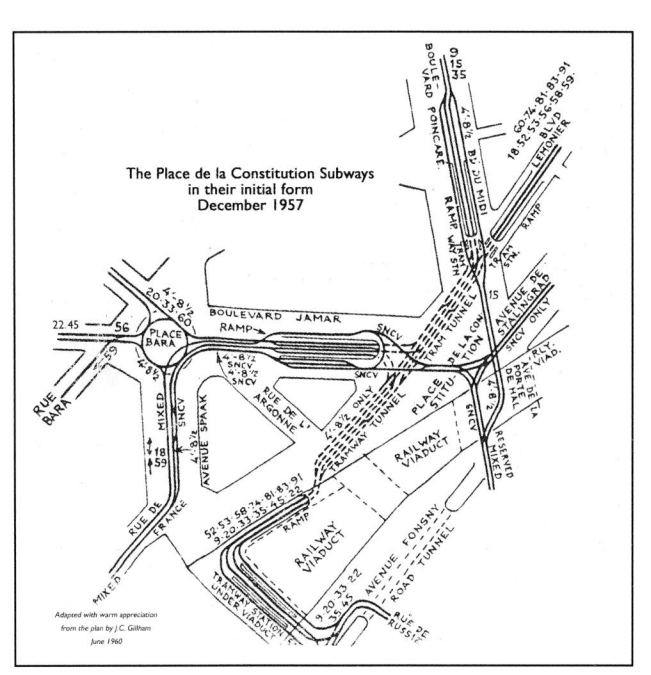

Map 8. The Midi subways in their 1957 state. [Courtesy the late John Gillham]

An inter-war view shows the grandiloquent frontage of the old Gare du Midi, with TB trams calling conveniently just outside. The structure on the right is the unfinished abutment of the Junction Railway as it was left when work ceased in 1914. The station building, so characteristic of the pride and flamboyance of the booming Belgian state, was completed in 1869. This site is near the location of the 'Rue Couverte' tram station below the present tracks.

The Place de la Constitution before the catastrophe of 1914, when it was a gracious and placid square. All this was swept away by the Junction works and by the rebuilding of the station: the Midi subways pass below the site today.

In their early days the Midi subways saw both traditional trams and the last word in PCC progress: two 7000-types swoop down the four track Jamar ramp.

We are now looking towards the Junction viaduct in 1961, with the Jamar ramp (and a STIB trailer) on the right, and a four-car Vicinal rush-hour train for Leerbeek on the left. These metre-gauge tracks were to have formed the Vicinal terminus, with through services to Place Rouppe ceasing, but public protest prevented this.

A fascinating scene from the Junction viaduct shows how Vicinal trains continued awkwardly to cross the entire road layout outside Midi station on their way to and from Avenue de Stalingrad (in the background) and their Place Rouppe terminus. The Poincaré branch of the STIB subway passes below this scene, with its ramp at the extreme left of the view, and Lemonnier station underneath

with concrete at their bases. The plan, as seen in the diagrams, is roughly shaped like an inclined letter 'K', the main stem forming the link between Boulevard Lemonnier to the north and the Midi station, the other limbs being the branches to Boulevard Jamar and Boulevard Poincaré respectively.

The track layout involved a four-track section (14m wide) through the main section of the tunnel and on the Jamar ramp: this enabled conflicting moves to be ameliorated by pre-sorting of trams. There was one station within the tunnel, at the north end. It had four platforms in the apex of the junction, each 4.2m wide and raised just above track level. The station had five entrances, two of which were originally equipped with escalators. The station walls were tiled and brightly lit, and the tram tunnels themselves were painted in bright reflective paint and lit throughout. The whole station complex would be called 'Lemonnier' after the completion of future line 3 as a pre-metro, but at this stage the two branching sections were separately named after the boulevards they accessed, 'Poincaré' and 'Lemonnier'. A simple stopping place, without shelters, was located on the Jamar ramp. The three original street ramps were finished in a distinctive and elegant fashion, with stylish fencing: two may still be seen today. In contrast the ramps on subsequent pre-metro lines were, in the main, makeshift and utilitarian, symptomatic of their intended temporary nature although some have lasted for decades. The initial stops used ground-level boarding requiring ascent to high-floor cars, but the low platforms were long enough to accommodate several trams at once. At this stage, of course, trolley current collection remained universal on all STIB and most SNCV lines in Brussels, and the initial subways were used by all classes of rolling stock, including wooden-bodied cars (see section 5.52 for details).

There were no signals in the tunnels, but compulsory stops were observed at pedestrian crossings between platforms.

There was also a four-platform tram station beneath the station itself with direct access from the concourse. There were originally subways to these platforms but passengers understandably preferred to take their chance on the level and these were soon sealed off. There was also a covered bus station to the north, mainly for the SNCB's complementary road services. Extensive reordering of tracks in the surrounding streets also provided new and improved facilities for the SNCV's metre gauge lines in the area. It had originally been intended that the Vicinal lines would be cut back from their terminal run towards Place Rouppe, and to this end an elaborate turning circle and storage sidings were built around the subway ramp in Boulevard Jamar. Public outcry at the enforced change by Vicinal passengers reversed this plan, and trams continued to run through, apart from some peak-hour extras. Of necessity these Vicinal cars running into the Avenue de Stalingrad still had to cross the Ceinture on the level.

For those of a certain age this image is painfully evocative of an era of smart clothes, tidy surroundings, and the new subways. We are looking up the Poincaré ramp in September 1958, with a motor and trailer set on their way towards Porte de Ninove. The two branches of the station had different names, indicating the surface streets to which they led. The original rather elegant art-deco décor is apparent, mostly gone now. The view also shows the low-level 'platforms', and the ample space for successive trams to draw up.

In 2018 the Poincaré branch platforms at Lemonnier show little of the style of sixty years ago, and improvement is blighted by uncertainty over the future layout.

Although much of the Midi subway is still in use this northern ramp towards Boulevard M. Lemonnier has gone. It was subsumed within the longer tunnel towards Gare du Nord in 1976 and the present station lies beneath this site. By-passing this location will be a major task in reconstructing the line for metro trains. [Courtesy John Bromley]

The southerly ramp of the subways was neatly contrived below the railway viaduct, where six-axle PCC No 7766 is climbing up to Rue Couverte in September 1987, on its way to Montgomery on service 81. On the left the fencing surrounds the massive works associated with construction of the new deep-level station, opened to trams down a new ramp in 1993.

On completion, the STIB tunnels were used by 3,566 trams each day, with a capacity of about 200,000 passengers.

At a small ceremony on 16th December 1957 the completed works were declared open, to be brought into public use the following morning. A bronze plaque in the apex of Lemonnier station, little noticed now, commemorates the event. Although much altered the basic elements of the 1957 tunnels and stations were still in use in 2018, although the impending conversion of the north-south tunnel to Metro use will destroy most of them. The original grooved tramway rail has been replaced by Vignoles rail and the Rue Couverte station has been reduced from four tracks to two, as has the Jamar ramp.

When the Midi tunnels were completed the ramp at the northern end, leading to Boulevard Lemonnier, was a temporary construction, and it was expected that the next stage of tunnelling would create a south-north link extending below the centre to the Nord station. In the event this was not the course which events took, and it took nine years before this challenging line was complete, as we shall see in a later chapter.

A breath-taking view from a train on the Junction viaduct shows the scene towards Bvd Poincaré and Porte de Ninove in the 1950s. In the centre is the ramp leading out of the subways, and just to the left of that are the two little pavilions covering stairways to the Poincaré branch of Lemonnier station. Just visible to the left of the ramp itself is a tram on the surface tracks which by-passed the station and the tunnels. The ramp is still in use today although the boulevard is much changed.

4.4 Wider Subway Plans from 1964

Writing in 1964 M Pierre Reynaert, General Manager of the STIB, set out clearly the need for public transport improvements. He pointed out that the growth in personal motor transport, whose users found public transport unacceptable because of its poor quality, was stifling traffic circulation and threatening economic progress. It was essential to restore standards of speed, regularity and dependability. Inevitably investment in a fixed track system offered the best prospects, but this would require concentration of services on to fewer lines and hence a simplification of the route network. Faster, high-capacity, vehicles were also required to enhance what would now (regrettably) be called 'the travel experience'.

Several possibilities presented themselves, ranging from an extensive full Metro (expensive and necessary only for flows of more than 20,000 passengers per direction per hour); elevated monorails or similar technologies (intrusive and unacceptable in central Brussels at least); and upgraded tramways with shallow subways where necessary, and segregated suburban tracks as extensively as possible elsewhere. He noted that many Brussels roads were broad enough to cater for such segregation and that an incremental process of shallow tunnel construction was practicable: this was his preferred solution.

Reynaert's specifications reflected plans for subway/surface operations in several countries, notably in Western Germany, although in Brussels itself (unlike in Antwerpen) it was a concept which faltered after 1976. Implicit in the whole idea, though, was the conviction, prevalent at the time, that the streets belonged primarily to private cars: public transport might be necessary, and there were arguments (as Reynaert said) for making it more attractive, but it needed to be kept in its place. Only with the upsurge in the 1980s in 'new tramways' in France and elsewhere was this doctrine finally put to rest, although this also effectively meant the end, in most cases, of further development of the tramway tunnel as conceived in the 'golden sixties'.

Before 1940 the focus of Brussels tramway operation was arguably the Bourse and the central markets. From the 1950s this changed, and new, more dispersed, activity required radical change in the provision of public transport. This can be seen in the developments we shall discuss later, serving the commercial area around Gare du Nord, NATO at Evere, the EU institutions around Schuman, and the repurposed Heysel Exhibition site.

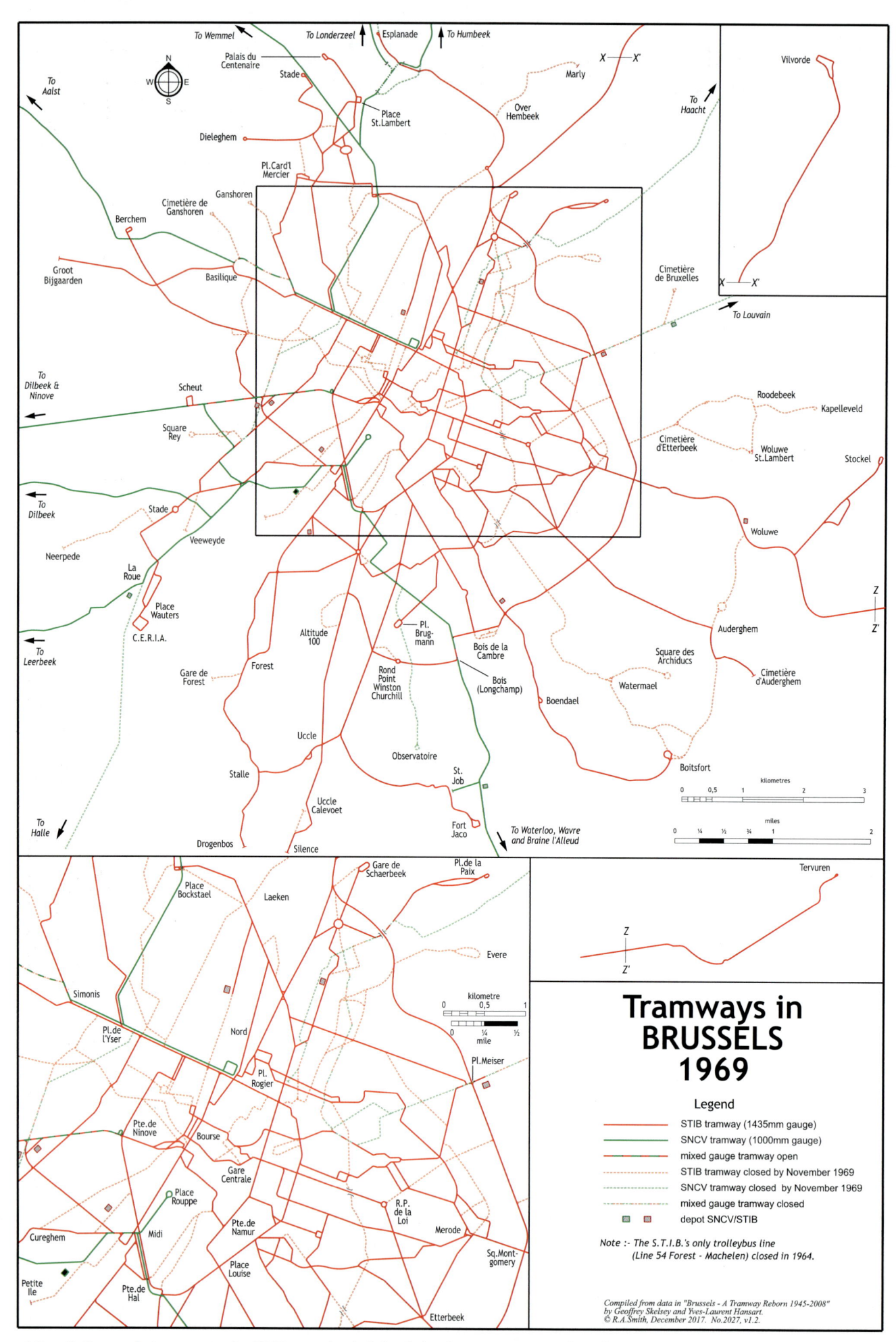

Map 9. Brussels tramways in 1969: on the brink of the first pre-metro. [RAS]

5. The Pre-Metro Arrives in Brussels

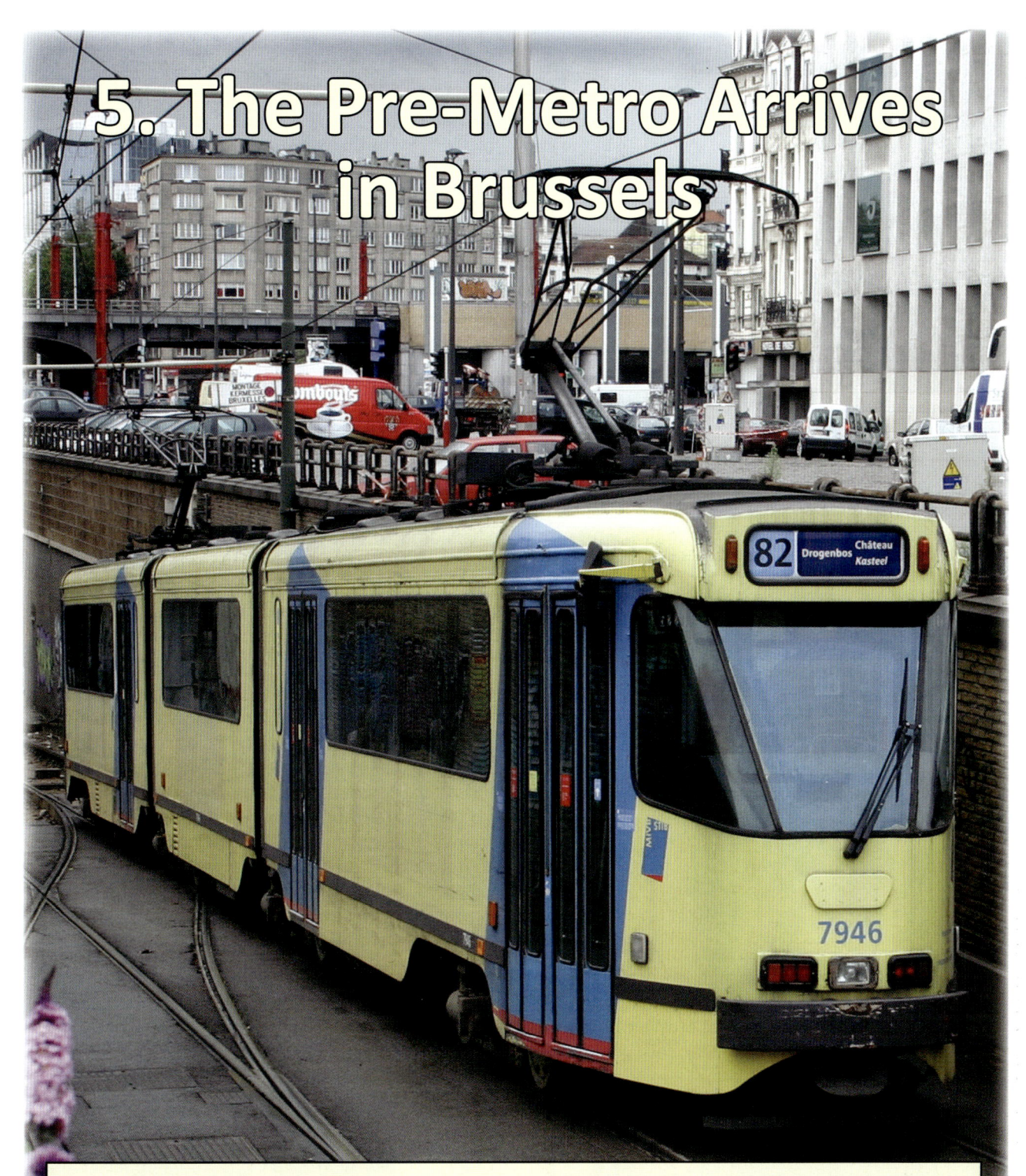

President J.F. Kennedy, in a message to Congress in 1962 commending a programme of federal capital assistance for mass transportation, expressed eloquently the case for rescuing American urban transport from the state of disgraceful decrepitude into which much of it had sunk:

"To conserve and enhance values in existing urban areas is essential. … Our national welfare therefore requires the provision of good urban transportation, with the properly balanced use of private vehicles and modern mass transport to help shape as well as serve urban growth."

The Urban Mass Transit Act of 1964 which followed provided $375 million for large-scale urban rail projects in the form of matching funds to cities and states. A Federal organisation, the Urban Mass Transportation Administration, was created to oversee this. In this chapter we see how Belgium moved in line with the spirit of the age.

The Midi tunnels showed how separation of tram and motor traffic delivered benefits for both. They became the blue print for the pre-metro concept: a 7900-type tram enters the Jamar portal, now reduced to double track.

5.1

As we begin our discussion of the extent and implementation of the Capital's underground system it will help first to list the opening of all the pre-metro sections. The line numbers listed are as used by the STIB in the 1980s (No. 4 was reserved for an unbuilt line, probably on the alignment of the Rue Royale). Map 10 shows the pattern of pre-metro construction envisaged in the centre at the start of the planning process, and Map 33 the completed system.

Line No.	Section	Length km	Stations	Opened as pre-metro	Converted to Metro
-	*Gare du Midi area*	*0.75*	*1*	*1957*	-
1	Schuman – Ste-Catherine	2.76	6	1969	1976
2	Namur – Madou (Petite Ceinture)	2.12	4	1970	1988
5	Meiser – Diamant (Grande Ceinture)	1.08	1	1972	-
2	Madou – Rogier (Petite Ceinture)	1.13	2	1974	1988
5	Diamant – Boileau (Grande Ceinture)	1.50	3	1975	-
-	Montgomery loop	0.20	1	1975	-
3	Nord – Lemonnier	2.97	6	1976	-
2	Louise station	0.55	1	1985	1988
-	Simonis station and approaches	0.50	1	1986	1988 (part)
3	Midi – Albert	2.99	4	1993	-

The first priority for elimination of surface tramways was the east-west transit of the old centre: this atmospheric view from 1952 shows a motor-and-trailer set emerging from narrow Rue Assault, with Ste-Gudule's church in the distance. All these lines had gone by 1970. [Courtesy Patrick Sellar]

RIGHT: Map 10. Tram subway and extension proposals, 1960-63. The STIB and the city drew up detailed plans for tramway improvements and these maps are a synthesis of the proposals. It will be seen later that some were fulfilled over the ensuing decades. [RAS]

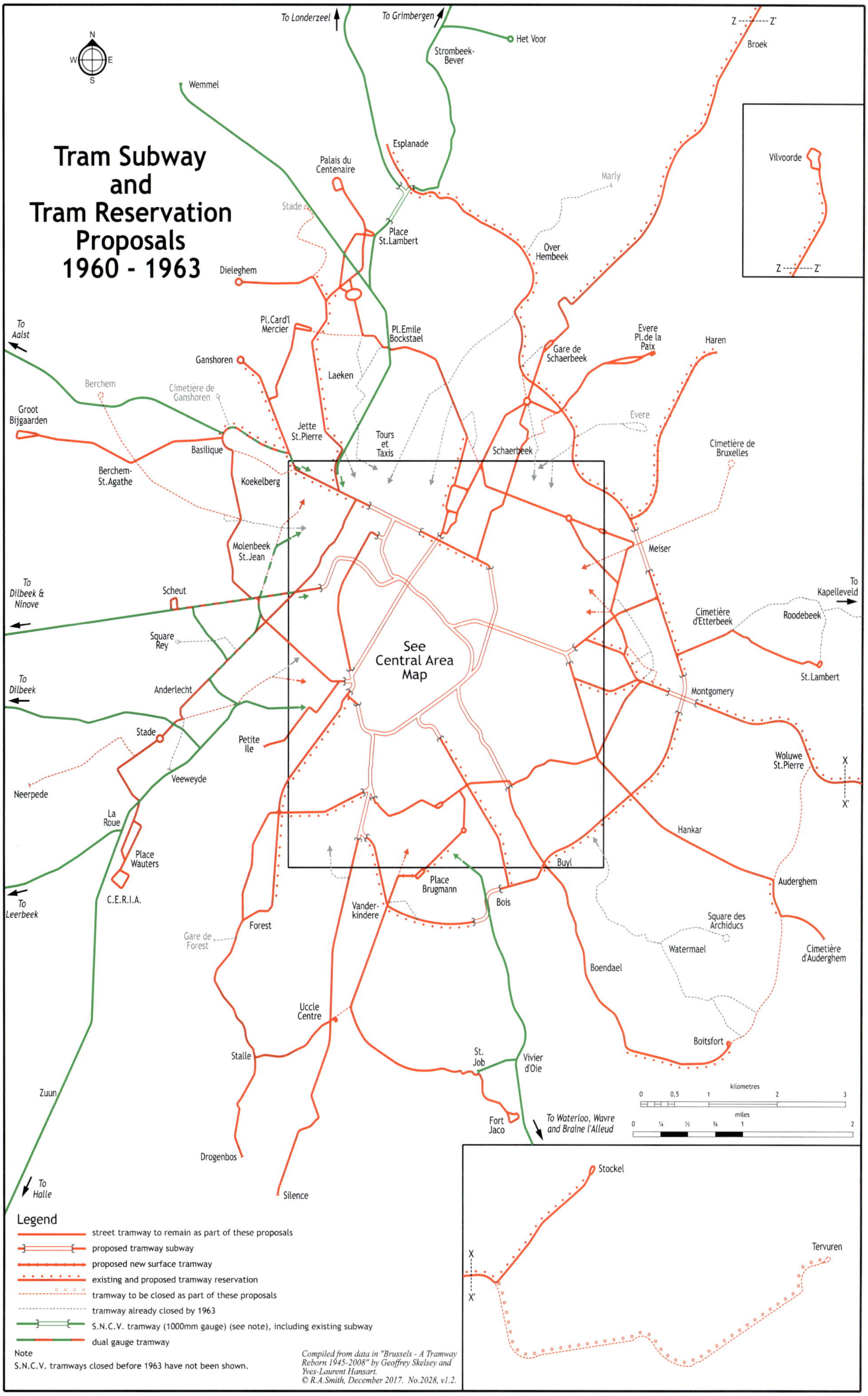
Tram Subway and Tram Reservation Proposals 1960 - 1963
N
W
E
S
To Londerzeel
To Grimbergen
Het Voor
Strombeek-Bever
Wemmel
Z - Z'
Broek
Vilvoorde
Z - Z'
Esplanade
Palais du Centenaire
Stade
Marly
Place St.Lambert
Over Hembeek
Dieleghem
To Aalst
Pl.Card'l Mercier
Pl.Emile Bockstael
Evere Pl.de la Paix
Haren
Gare de Schaerbeek
Ganshoren
Laeken
Berchem
Cimetière de Ganshoren
Groot Bijgaarden
Evere
Jette St.Pierre
Tours et Taxis
Basilique
Schaerbeek
Cimetière de Bruxelles
Berchem-St.Agathe
Koekelberg
Meiser
Molenbeek St.Jean
To Dilbeek & Ninove
Scheut
To Kapelleveld
Cimetière d'Etterbeek
Roodebeek
Square Rey
See Central Area Map
St.Lambert
To Dilbeek
Anderlecht
Montgomery
Stade
Petite Ile
Woluwe St.Pierre
X
X'
Neerpede
Veeweyde
La Roue
Hankar
Place Wauters
Buyl
C.E.R.I.A.
Place Brugmann
Auderghem
To Leerbeek
Bois
Vanderkindere
Square des Archiducs
Forest
Gare de Forest
Watermael
Cimetière d'Auderghem
Boendael
Uccle Centre
Boitsfort
Stalle
St. Job
Vivier d'Oie
kilometres
0
0,5
1
2
3
Zuun
miles
0
¼
½
¾
1
2
Fort Jaco
To Waterloo, Wavre and Braine l'Alleud
Drogenbos
Stockel
To Halle
Silence
Tervuren
X
X'
Legend
street tramway to remain as part of these proposals
proposed tramway subway
proposed new surface tramway
existing and proposed tramway reservation
tramway to be closed as part of these proposals
tramway already closed by 1963
S.N.C.V. tramway (1000mm gauge) (see note), including existing subway
dual gauge tramway
Note
S.N.C.V. tramways closed before 1963 have not been shown.
Compiled from data in "Brussels - A Tramway Reborn 1945-2008" by Geoffrey Skelsey and Yves-Laurent Hansart.
© R.A.Smith, December 2017. No.2028, v1.2.

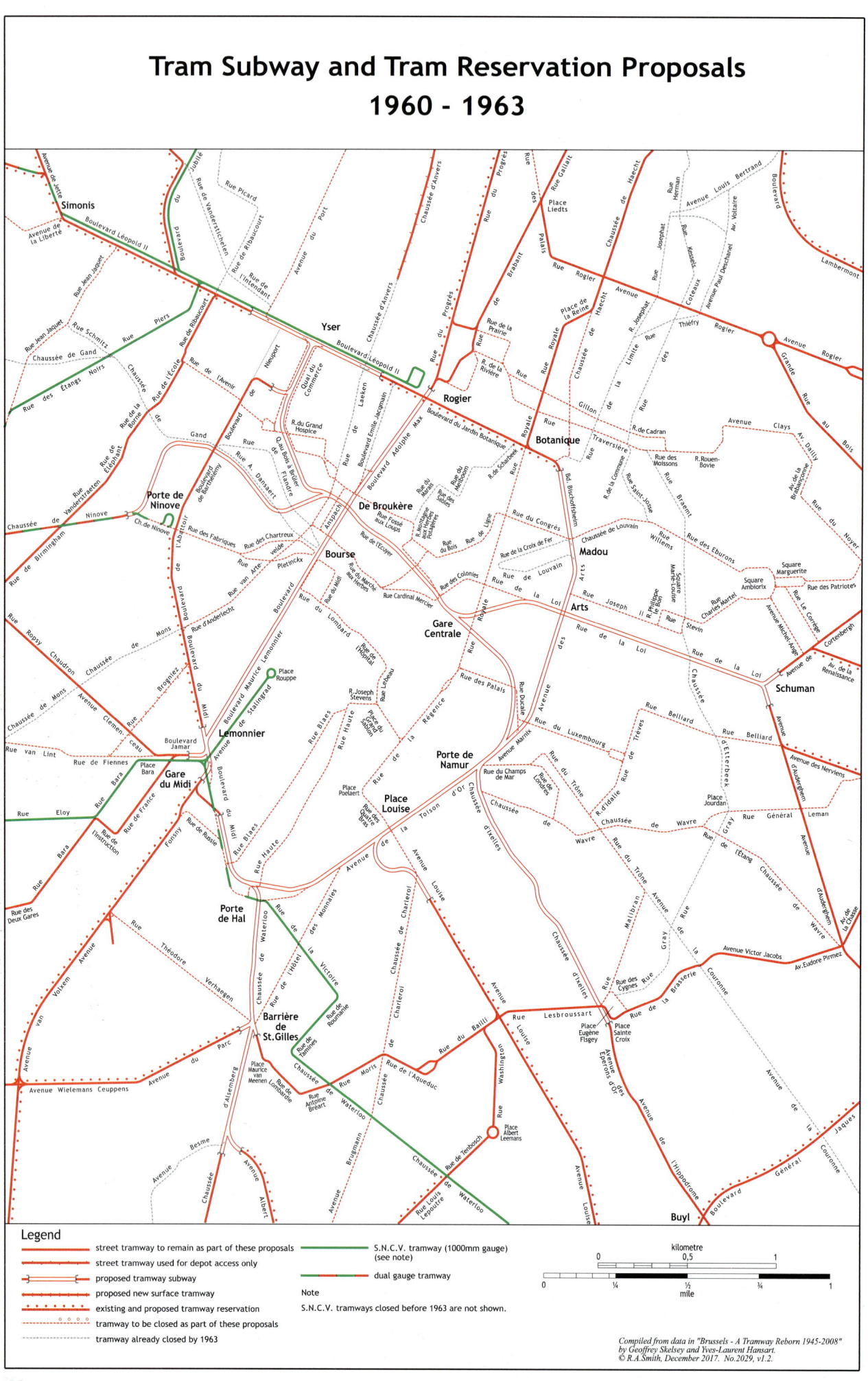

Tram Subway and Tram Reservation Proposals 1960 - 1963

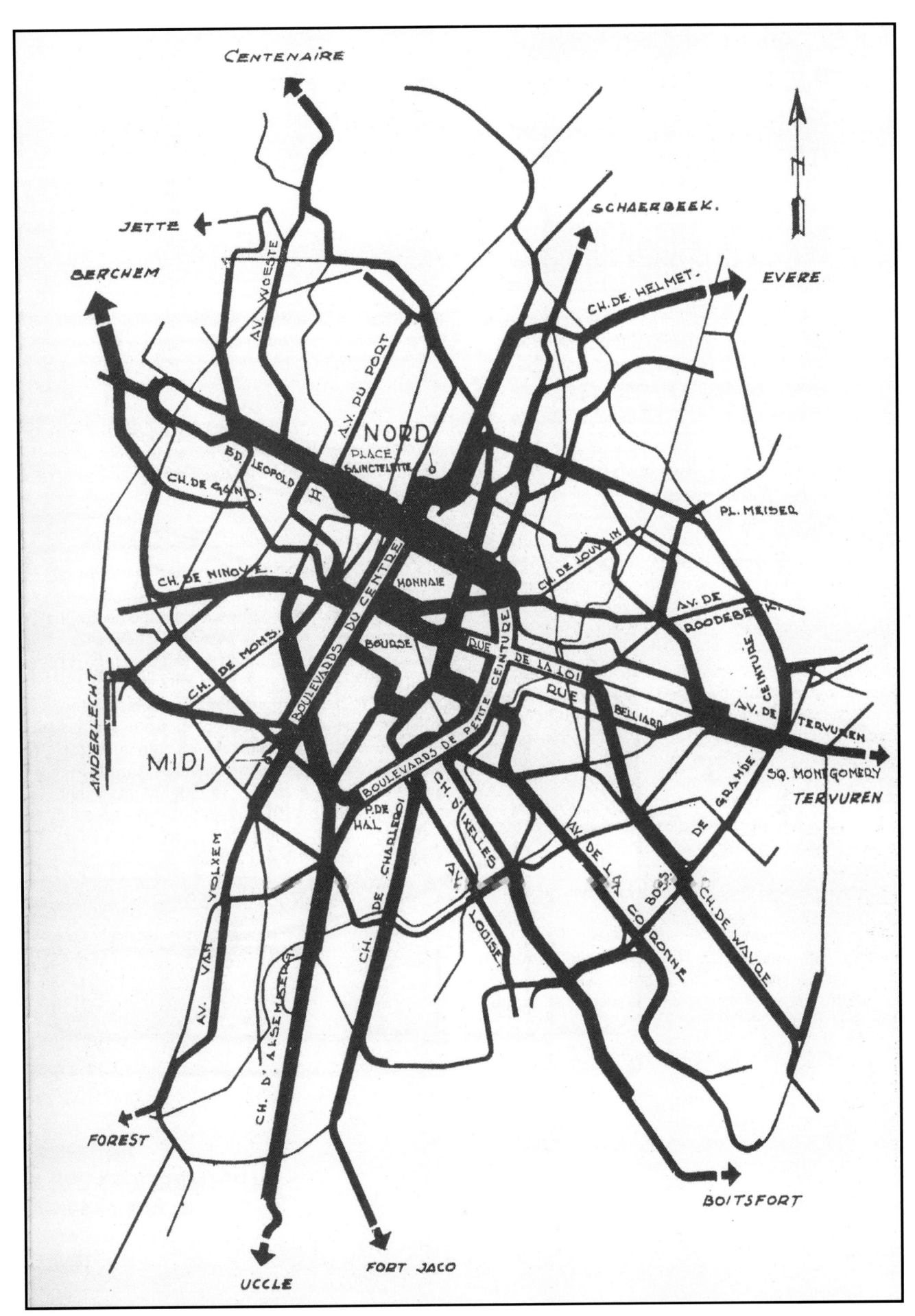

LEFT: Map 11. Central area subway proposals 1960-63. [RAS]
ABOVE: Map 12. The STIB carried out surveys of peak-hour traffic densities in winter 1963-64. The resulting plans identified sections which justified upgrading. [Authors' collection]

5.2 National and Regional Studies

In January 1962 the Belgian Government set up a permanent central organization within the Ministry of Communications (i.e., Transport and Telegraphs), *La Service de Promotion des Transports Urbains* (PTU) to coordinate a programme of improvements to urban mass transit. Each of the principal urban areas – Brussels, Gent, Antwerpen, Liège, and Charleroi - was charged with establishing a local body (*Service Special des Etudes*, SSE; or *Bijzondere Studiedienst*, BS) to draw up and oversee their local studies and their implementation.

The Brussels SSE began its work in 1963, starting with a wide-ranging survey of traffic and congestion to establish areas most in need of improvement. It was found, unsurprisingly, that the Grands Boulevards between the Nord and Midi stations, the east-west axis between the old city and Schuman, and the eastern side of the Petite Ceinture were the least satisfactory areas for current public transport, with a number of isolated pinch-points elsewhere. A survey of existing transport provision (see map 12) showed that by far the densest tramway traffic coincided with these axes. There followed in 1964 an assessment of the new construction necessary to address the problem, the cost of which would under the new national arrangements be largely met from central funds. Map 10 shows an early formulation of what became the Brussels pre-metro, and later Metro, systems. This envisaged radical new roles for tramways: surface lines would disappear entirely from within the 'Pentagon', succeeded by a coarser net consisting of 18.6km of peripheral, north-south, and east-west axes much as eventually came into effect. At this stage, as seen on the map, the three elements were not intended to be separate entities, with passengers having to walk between them at interchange points: instead all three were to be joined by running connections (although even with grade-separated junctions, as envisaged, it is questionable if so many closely-spaced intersections would have been operationally practicable). This would have introduced considerable flexibility into the system, in theory permitting varied service groups as successors to the complex tramway routes.

Other tunnels were proposed below Rue de la Loi, Chaussée d'Ixelles, and Chaussée d'Alsemberg, with shorter local underpasses elsewhere, such as at Place Meiser, Montgomery, and Bois de la Cambre. However, as was clear from the density diagram some parts of the net of existing tramways carried traffic far below that needed to justify investment in fixed track transport and a systematic weeding-out of such lines in favour of buses soon followed. A significant tramway network was to be retained, feeding into the new tunnels, and work was to be undertaken to transfer as many of these lines as possible to segregated track. The long-term network resembles in part the tram system as it is today, although few suburban extensions were then proposed.

The outcome of the studies was an enviable long-term policy to produce a city-wide high-capacity transport system. The rolling programme of tunnelling was planned to run at a consistent investment rate of around 1000 million Belgian Francs a year (about £7.44 million at 1964 prices, although inflation has robbed these figures of any relevance); in fact construction continued with breaks from 1965 until 2009, and is set to resume in 2019.

5.3 Why Pre-Metro?

An official Belgian publication describes the concept of the 'pre-metro' as 'a Brussels specificity'. This may not be strictly true, as we have seen several other contemporary applications, but it is the case that the extended period of tramway operation, and gradual substitution of metro trains over four decades, became the practice in Brussels. (The similar project in Antwerpen was never so converted and remains wholly tram worked). During the early 1960s STIB officials made contact with twelve other tramway undertakings contemplating similar projects and these studies, together with experience of other modern metro systems, informed the planning process.

The 1963 Brussels studies concluded that both practical policy and economy suggested that construction of the new network should be gradual and incremental, with each tunnel section, as it was completed, used in the first instance for a 'pre-metro' service of tramcars, and only ultimately upgraded to heavier operation. The infrastructure would be designed and built for metro use, although temporarily fitted out for tramway standards, for example in respect of platform height and current collection. This policy had a number of important advantages. For one thing it was possible gradually to develop a high capacity transport system with city centre tunnels linked to upgraded suburban tramways, so that the benefit of investment was experienced sooner and more widely. For another,

"Up a steep and very narrow stairway": accessing 7000-types was difficult'.

Typical of scenes across the growing pre-metro network is this view inside Luxembourg (later 'Trone') station on line 2 in 1970, as a 7000-type tram makes a trial run and workers put finishing touches to the structure. Notice the lowered sections of the platforms, the original lettering style and décor, the train indicator equipment, and in the foreground the contactor for the automatic braking system.

it was possible to retain high quality, high-capacity transport over each section until a brief period of suspension during conversion. This policy took effect in Brussels between 1969 and 1988. Finally, lengths of tunnel could be brought into operation successively as they were completed, using temporary ramps at the completion of each phase of construction, thus bringing progressive relief from surface traffic congestion. The sequential opening and up-grading of the 'inner ring' subway in five stages between 1970 and 1988 exemplifies this concept, as described later. Such eventual upgrading of tram subways to metro operation was widely advocated at this period, but conversion of Line 1 of the Brussels system was the first after 1950 to actually take effect.

There was however a flaw in the pre-metro argument, which could have been addressed but wasn't in Belgium. The nature of Brussels tramway rolling stock at the period, and in particular its slow loading and unloading through narrow doors with steep steps, together with the lack of multiple-unit equipment, meant that the expensive subways and stations could not be fully utilised with conventional tramway operation; this severely limited the capacity of the services. Such issues were being addressed at exactly this period, notably in Frankfurt-am-Main where hybrid 'tram/metro' rolling stock was tested in 1964 (and shown at the München exhibition that year). Its successors are still operating today in Frankfurt. No such step was contemplated in Brussels until 1978, and then in vain. It was not to be until the twenty-first century that access and capacity issues were finally dealt with on Brussels tramways, and by then the availability of higher-capacity low-floor trams and the further simplification of the tramway service pattern made possible a true 'métro-léger' or light rail system.

An interesting pair of pictures shows one of the portals west of Schuman station under construction, and in use. This is the Joyeuse Entrée portal on the south side, giving access to Avenue d'Auderghem and in use from 1970 to 1976. The second portal to the right was for a road tunnel, still in use today.
The second view shows the portal towards the end of its life in April 1976, with a Tervuren-bound car leaving the subway. For the last year before the line was converted to metro operation the northern 'Rennaissance' portal was taken out of use and all services used this ramp. [MJR]

43

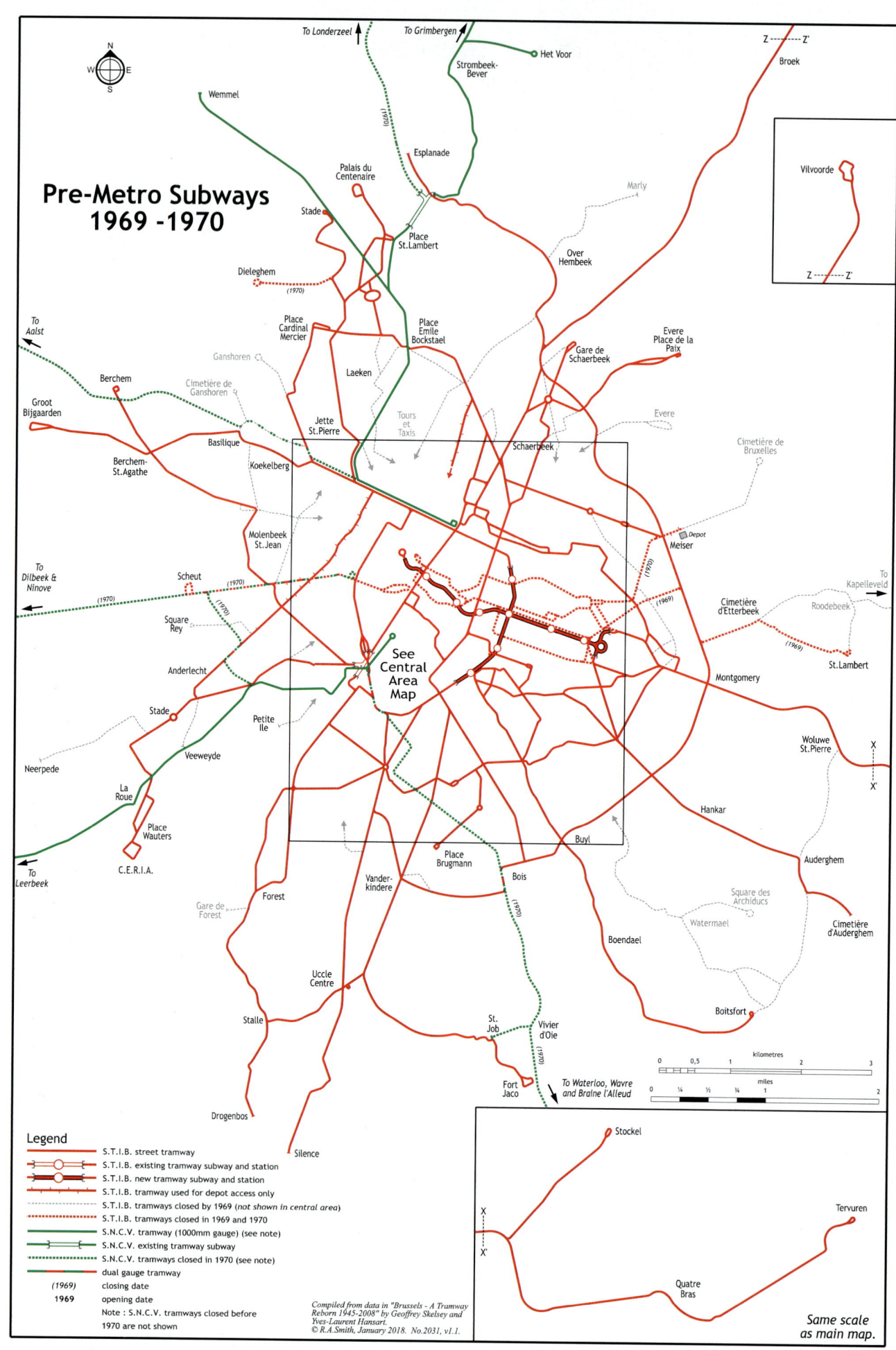

Map 13. Pre-metro developments 1969-70 [RAS]

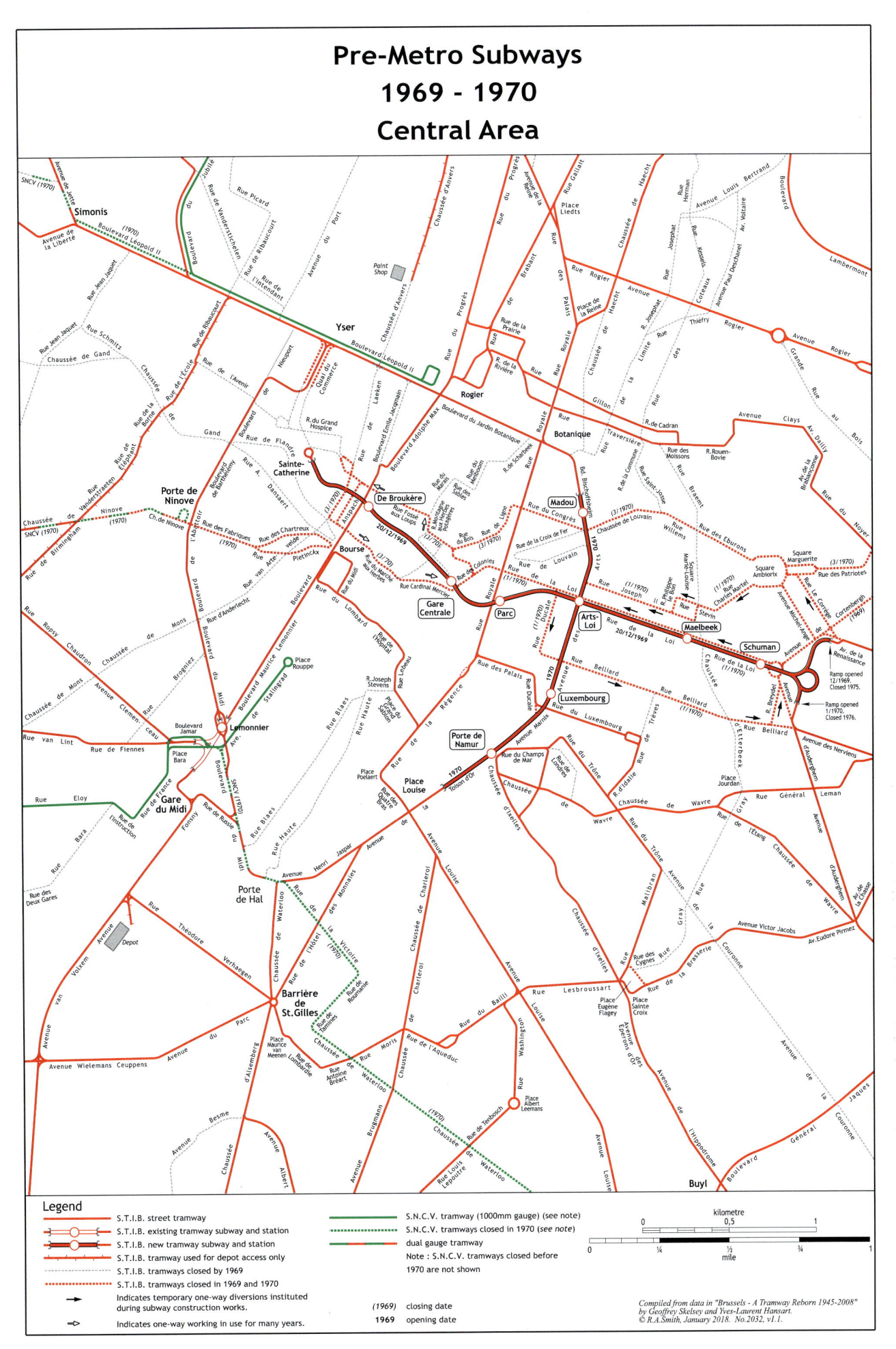

Map 14. Central area pre-metro, 1969-70 [RAS]

5.4 The Achievement
We will now look at the four pre-metro sections completed before 1988, in partial fulfilment of the plans developed from 1965.

5.41 Line 1 (East-West)

The massive nineteenth century re-ordering of much of the urban fabric of Brussels left an enviable legacy of broad roads, notably along the line of the former fortifications ('*La Petite Ceinture*'), radial roads such as the Avenue Louise, and the straight new road between the Midi and Nord stations ('*Boulevard du Centre*'). But the complicated tangle of ancient thoroughfares around the Grand' Place, at the heart of the old city, resisted even the determination of successive kings, and no east-west road was built to match the spacious north-south and circumferential boulevards. Unsurprisingly therefore, this important axis, handling heavy tramway traffic, was plagued by congestion and delay and was a priority for underground relief. Although, as we have seen, the first such line to be started was to have been the north-south axis other developments determined the actual order of construction. The planning of what became the 'European Quarter' along Rue de la Loi required construction of new road tunnels there and below the Parc du Cinquantenaire, and for engineering reasons it made sense to combine those works with rail construction nearby and so to commence with what became Line 1.

The first excavation for the east-west line was started at Rond-Point Schuman in May 1965. Experienced tunnellers had been recruited from closing coal mines in Limbourg and Wallonie, drawing on skills which would otherwise have been dispersed. All the new tunnels were designed from the start, in dimensions, curvature, and profile, for eventual use by metro trains, unlike the 1957 tram subways at Midi where the original tramway standards are still visible today, with steeper ramps, sharper curves, flat junctions, and narrower passenger platforms.

The decision largely to use cut-and-cover construction made necessary an elaborate series of surface track diversions, to maintain tram services during excavations. Lengths of new tramway were opened in June 1966 to allow one-way working around construction sites in Rue de la Loi. Later in 1966 a further diversion was built around the site of the road entrance ramp near the future Schuman underground station. Similar tramway diversions were introduced in 1967 within the Pentagon, to avoid the tunnelling work beneath Rue de Loxum and Rue d'Arenberg near the Central Station, a particularly difficult construction site.

Much of the initial line was excavated beneath streets. Although construction of tunnels using large open pits was employed in places, a preferred technique where site conditions allowed was to excavate two parallel trenches as far apart as the width of the eventual tunnel. These were then filled with reinforced concrete in situ, and eventually

This is Rue de la Loi in 1967, showing the street still in partial use whilst beams are inserted at the outer edge of the future tunnel.

Building line 1 was a daunting task with several difficult sections. East of Parc it was necessary to use a boring machine, producing a circular 'tube' tunnel.

The scale of some station construction was remarkable, as at Schuman in 1967.

The pre-metro in operation: a single-ended six axle car calls at Maelbeek station on 19 April 1976, not long before conversion: the third rail for metro trains has already been installed and preliminary work is taking place to infill the lowered platforms. Notice the mirror to allow the motorman to check the doors: the red target on the dash shows that this is a one-person-operated car. [MJR]

Representative of the long-drawn-out process of construction is this scene near Porte de Tervuren, showing work in May 1972 on excavation of the eastern prolongation of line 1, with the temporary road viaduct on the right. A 5000-type bogie car heads west on temporary track. [MJR]

With Porte de Tervuren (Merode) and Cinquantenaire in the distance this car heading for Montgomery is on the surviving reserved track alongside Avenue de Tervuren. Line 1 metro is underneath and to the left is the road tunnel passing under the Square. This section required massive excavations.

formed the side walls of the running tunnels. They were decked over during excavation thus preserving passage along the street above, and although the works were still intrusive there was less interruption to daily traffic. Between Arts-Loi and Parc stations a tunnelling shield was used, and elsewhere chemical consolidation of the ground and other techniques were required. Most construction was through sand, but with particular difficulties where the line passed below the valley of the Maelbeek, where diaphragm or 'slurry' walls were required for some distance through waterlogged ground. The cut-and-cover tunnels were rectangular in cross-section, 7.4m wide and 4.5m high. The circular bored tunnels had a diameter of 8.3m. All running tunnels contained double ballasted tracks, with rails laid on hardwood sleepers. There were narrow walkways on either side, which also contained water mains for fire-fighting and other purposes. An unusual intended feature was the incorporation of pneumatic tubes for the Belgian Post Office's mail and telegram despatch system, although this was never in fact activated.

Signalling employed small coloured light equipment, supplemented by panels indicating one of a range of mandatory speed limits. Provision existed for a second tram to enter an occupied platform, to maximise throughput. Operation was monitored from a comprehensive control centre located at Parc station, and closed-circuit television was used to oversee stations. Radio communication was provided between controllers and drivers, using 'leaky' coaxial cables through the tunnels.

All trams using the tunnels were fitted with a form of automatic train- and speed- control, and there was a vehicle identification system based on American 'Syvania' equipment, which used reflective strips on vehicles 'read' by photo-electric devices. These also actuated passenger information systems on the platforms, showing the service numbers of the next three departures. For compatibility with tramway operation overhead power supply was required during the pre-metro phase. It was of course necessary for trams working through the tunnels to use pantograph collectors, but as the overhead line on the surface system had yet to be converted semi-automatic change-over devices were required at portals for transfer between collectors.

There were six underground stations (De Brouckère, Gare Centrale, Parc (originally to be called 'Royale'), Arts (renamed 'Arts-Loi' from 11th November 1972), Maelbeek, and Schuman).

The most obvious feature of the stations during the pre-metro phase was the configuration of the platforms, each of which included a short lower section enabling boarding of tramcars from near track level. This can still be seen today at stations on lines 3 and 5. They were designed to be raised to form continuous high-level platforms at the time of conversion. The platforms are, however, somewhat narrower than those built later on Metro extensions.

At its eastern end trams on pre-metro line 1 emerged from the tunnel into a spacious loop terminus at Ste-Catherine, in use for only seven years. This west-bound car for Stockel is just entering the tunnel on 19 April 1976 on the way to De Brouckère, and just above it is the supervisory control cabin. After closure to trams a subway station replaced this terminus. [MJR]

Compared with some of those which came later, the initial stations were relatively simple but fitted out with some style and with relatively high ceilings contrasting greatly with older installations in London, Paris, and New York. Wall coverings were in marble, and advertising was restrained. Mosaic was widely used in access passages and there was generous provision of escalators. Signage was carefully thought out, using a special type face designed by the celebrated Belgian typographer Fernand Baudin: a new visual identity was created in the twenty-first century. Station names were, of course, alternated in both national languages, and pictograms were widely used. A new, if somewhat eccentric, metro 'identifier' was adopted nationally and is still used in Brussels and Charleroi.

On Wednesday 17th December 1969, using specially-prepared PCC tram 7093 (now preserved), H.M. King Baudouin formally inaugurated the first 4.2 km of underground tramway, which ran between a ramp east of Ste-Catherine and a point east of Rond Point Schuman. The King began his journey at De Brouckère, accompanied by a sizeable suite including the Mayor, Prime Minister, Ministers, and business and civic leaders. At Schuman His Majesty was shown features of the station and of the line's construction, after which for three days there was free access for visitors to inspect the new stations in groups of fifty. Stress was laid on the need for pre-payment using ticket machines, and of gated access to platforms (which was subsequently discontinued until recent times). 108,000 visitors were admitted between 17th and 19th December. Public operation began on Saturday 20th December. On that day the SNCB also opened a new interchange station at Schuman, which would assume growing importance with the growth of employment in the nearby 'European' quarter. This was the beginning of a slow move towards dedicated intermodal links.

The opening of the tunnelled section radically altered the pattern of east-west surface tram traffic. A new surface turning loop and sidings were constructed in the former Fish Market at Ste-Catherine, where the subway routes emerged from the tunnel: no tram services proceeded further west of here, the ground having been prepared in a 1967-8 restructuring by converting the western branches to bus operation. At the eastern end twin ramps were built on branches just beyond Schuman station, at Avenue de la Renaissance and Avenue de la Joyeuse Entrée. These gave access via Avenue de la Renaissance and Avenue des Nerviens respectively to Cinquantenaire (Merode) and beyond. (Until January 1970 the Joyeuse Entrée ramp was incomplete and its sole user, service 23, made a detour via Cinquantenaire). Initially only services 23 (Bourse – Schaerbeek station via Montgomery), 39 (Bourse - Stockel) and 44 (Bourse - Tervuren) were diverted through the tunnel, and few surface tracks were abandoned, with the remaining tram services on this axis continuing on the street tracks for a few further weeks. An additional all-subway service ran between Ste-Catherine and an underground turning loop at Schuman, in the divergence leading to the two eastern portals. (Signs of these short-lived structures are still visible to keen-eyed travellers).

After a period of adjustment other services joined the three in the tunnel: from 10th January 1970, service 24 (Bourse - Bois) and 25 (Bourse - Auderghem) also began to use the subway. The whole of the once-heavy tramway traffic along the Rue de la Loi axis had disappeared underground. This left only two remaining services entering the Pentagon from this direction -the last surface tram routes crossing the centre between west and east suburbs- and these in turn were replaced by buses in March 1970. This finally made possible abandonment of the last of the surface tracks between Rue Royale and Bourse, a historic change as these serpentine routes along narrow streets, many used in one direction only, had been a feature of the electric tram system since its earliest days over seventy years before.

5.42 Line 2 ('Ceinture', Inner Circle)

The *'Petite Ceinture'*, most particularly its northern and eastern sectors, carried the densest tramway traffic on the system and was one of the great tramway arteries of Europe, resembling the similar 'Ring' lines in Wien. Fourteen different services used one segment or another of the eastern ring, with a tram passing every few seconds. The northern section had been improved in the 1950s, in connection with road improvements for the Exhibition, and a massive elevated viaduct stretched west from Place Rogier, with pairs of tramway tracks at ground level on either side of the supporting columns below its lips. On the eastern side of the Ceinture route tracks had gradually been relocated onto lateral reservation on the outer side of the highway, whilst road underpasses at several intersections took through road traffic away from the public transport lanes. Nonetheless congestion persisted as ten other tramway lines joined or crossed the Ceinture along these sections, and numerous light-controlled cross-streets further delayed tramway traffic. Some work was undertaken in installing improved track layouts, for instance so that trams turning left did not delay cars behind them, but these were palliative measures. A further complication in the 1960s was the presence on the northern sector and in the south-east of the metre gauge tracks of the Vicinal, and although a new four track layout operated between Place d'Yser and Nord from 1957, some conflicts persisted. Placing tracks underground was the only effective solution.

Work began on the 'line 2' tunnel in May 1965, when the then Minister of Communications, M A. Bertrand, started a pile-driver close to the future northern entry ramp to the tunnel at Square Henri Frick near Place Madou in the north-east. Stage I of the project covered the 2.1km section to Rue des Drapiers, near Place Louise, although this was to take over five years to complete, and it was to be twenty-three years before the initial project was completed in 1988 (and a further twenty for a full circuit to be built). Maintenance of tramway tracks again involved temporary relocations, and the construction of steel-supported structures whilst excavation proceeded underneath.

The Ceinture tunnel was opened in sectors with successive temporary ramps opened as each part was completed. The 2.1km between Madou and Avenue de la Toison d'Or was opened by H.R.H. Prince Albert (later King) on 30th March 1971. Four new stations were included, at Madou, Arts (later 'Arts-Loi'), Luxembourg (later 'Trône'), and Porte de Namur. Even this short section, amounting to about a fifth of the total intended tunnel, avoided several cross streets and improved reliability. On 11th December 1973 a surface tram station, with controlled entry and payment facilities, was opened at Place Louise just south of the temporary ramp. Work continued below the tramway both south of Louise (where a complex underground structure was being built, including passive provision for a future cross tunnel), and between Madou, Rogier, and Simonis.

On line 2 a nodal point was Place Rogier, where provision was made for future line 3. Before the tunnels were extended here in 1974 tracks were laid on roadside reservation, seen here looking towards Jardin Botanique. [MJR]

Just west of Rogier the Vicinal and STIB tracks drew together under the new road viaduct. Here an inbound N-type car from Grimbergen passes an outbound PCC. Behind the camera the two pairs of tracks take up their location on either side of the supporting columns. The cross tracks here were disused by 1963.

By 1984, further west along Bvd Bauduoin, work was in hand to replace the 1950s viaduct with new road and rail tunnels. This is Jacqmain with the STIB tram tracks relaid on the north side and a temporary structure replacing the former concrete viaduct. Double-ended PCC 7814 heads west.

49

Looking west from Jacqmain the ugly and noisy steel road viaduct is on the left as 7900-type car 7928 heads inbound. Completed after over five years work the tunnels here were never used by trams but converted directly to metro in 1988.

After 1970 Ceinture tunnels extended from Madou as far south as the approach to Place Louise. Shortly before completion this is the scene at the temporary ramp south of Porte de Namur, with temporary surface tracks by-passing the ramp which remained in use until 1985. [Courtesy John Bromley]

Luxembourg (now 'Trone') station on line 1 just before opening in 1970, on the first section of the Petite Ceinture pre-metro to be completed. Notice the 'close-up' signal on the right, allowing a second car to approach an occupied berth. The tram is of course operating with a pantograph in the tunnel, but retains its trolley-pole. Visible on the dash of the tram is the 'T' symbol, indicating a car equipped with pre-metro automatic braking equipment.

An unusual feature at Place Louise was a controlled-access surface 'station' for trams, installed to speed boarding by requiring pre-payment. These temporary arrangements are seen on a misty spring day in 1984. In the distance a tram is seen on the deviation tracks built around the excavations for the future Louise station, opened a year later.

The 1970 section of the Ceinture pre-metro ended at its northern end at Place Madou, pending completion in 1974 of the extensive construction works at Botanique and Rogier. This is the temporary Madou ramp looking south towards Arts-Loi station.

This is the Louise ramp again, seen in April 1984 looking north as a 7900-type eight axle car heads south. The portal was removed in 1985 when the tunnels were extended south, and by this time the tracks looked distinctly uneven.

Louise station and a further short southwards extension of the tunnel were completed in 1985, and this is the new portal in the final stages of construction (see also the view on the cover). This remained in use until 1988. [MJR]

A poor-quality view inside Louise pre-metro station in 1987 not long before conversion to metro operation. The lowered platform sections and overhead wires are clear: evidence of the latter lasted until recently, though the station has been renovated. Notice also the steep incline down towards Hotel des Monnaies in the distance. The mountings for the third rail have already been installed and work is proceeding on raising the platforms.

An informative view at Louise in 1984 looking across the Petite Ceinture towards Avenue Louise in the distance, a year before the new underground section here was completed. There were then curves joining the north-south and east-west routes, used by a number of useful services, but these were necessarily withdrawn when the pre-metro was complete. Comparison with the image on page 8 will show that the form of the Place, and some of the buildings, survived from seventy years earlier. The tracks away from the camera are still in use in 2018.

A particular complexity on the northern sector of the line was the presence of the 1950s-built viaduct above the road. Tunnelling here had originally been ruled out as too expensive and difficult by the existence of numerous watercourses, including the important Charleroi Canal, but it was now intended to build parallel rail and road tunnels throughout. The road viaduct had to be replaced piecemeal with temporary steel structures spanning the whole working site, as the existing road supported by central columns was removed section by section. This ugly and noisy structure remained in use for several years. During construction the tramway was progressively removed from its central position below the old viaduct to lateral reservation on the north side, with a temporary turning circle at Jacqmain to accommodate single-ended cars. Sadly, but conveniently from the point of view of the engineers, it was no longer necessary to cater for the metre gauge as the last Vicinal services were withdrawn in 1978. At that time it was intended that the last two busy routes (Wemmel and Grimbergen) would be converted to standard gauge and incorporated into the STIB but this has not occurred, partly because of the political complexities of extending STIB services beyond the regional boundary. (As we shall see the Vicinal subway at Heysel was however reused).

The next sector to be opened along the Petite Ceinture, on 19th August 1974, was the 1.13km stretch between Madou and Rogier, with stations at Botanique and Rogier. The latter station, below the Place, was another complex structure on three levels, including provision for future line 3. Further construction southwards took many years and a short 550 m extension, including a new underground Louise station, was not opened until 20th August 1985, with a temporary ramp further south near the Porte de Hal. Construction alongside this ancient structure, and immense complexities around Midi where a huge deep level construction was required with careful attention to the stability of the railway viaduct above, were significant factors in slowing the work. Trams were diverted into this short new section and underground station, and this meant that the former surface tramway connections in Place Louise joining the Petite Ceinture to the intersecting east-west tramway were abandoned, forcing reordering of tramway services.

A year later attention moved to the extreme west of the northern sector of the Ceinture, with a short section of subway and a new station opening on 23rd June 1986 at Simonis, with two new ramps to the west of the tramway station and an approach ramp from the surface tramway in the east. At this stage, it was assumed that the two remaining Ceinture sectors (between Louise and Midi, and between Rogier and Simonis) would be completed for tramway operation towards the end of the 1980s, but a sudden ministerial announcement in November 1986 brought forward the inauguration of metro service over the whole line, with the conversion of existing sections from pre-metro. We shall see later how this evolved.

South of the Louise portal reserved tracks remained alongside the Ceinture most of the way to Midi, and a southbound 7900-type is seen passing Porte de Hal in 1987. The 'circle' service became metro in 1988, but between here and Midi surface trams continued until completion of the line 3 tunnels in 1993.

We have seen the major works west of Rogier as road and rail tunnels were built to replace the 1950s viaduct. This is the scene after completion, but before the metro was opened in 1988. Temporary tracks are laid on the north side of the highway, which has been resurfaced. The vast Basilique church can be seen in the far distance, beyond Simonis where the rail tunnels ended. [MJR]

Entrances to the new pre-metro stations were originally indicated by these distinctive signs. This is Madou on the Ceinture, and the stairway is still in use. [Courtesy LRTA Collection/Online Transport Archive]

5.43 Line 3 (North-South) and Extensions

The North-South line followed the course of the Boulevards constructed in the 1870s, as part of a gigantic project to provide new trunk sewers and storm water drains, replacing the dangerously polluted and sinuous River Senne which had followed a similar course. The cost of this huge project was met by construction of high quality commercial space fronting the new streets. Great vaulted tunnels were built, with the new streets (eventually Boulevards Lemonnier, Anspach, and Adolph Max) above them. This was perhaps the city's premier tramway with ten or more different services in the 1950s, again subject to delay because of conflicting tram and motor traffic. It served not only residential and business premises along the Boulevards, but also the Stock Exchange, the Opera House, the entertainment district around Place de Brouckère, and leading hotels there and facing Place Rogier. The route's importance as a direct link between the Midi and Nord stations of course diminished with completion of the 'Junction' railway, but it was still an artery for those arriving at either station with business in the capital's commercial centre.

As we have seen, this was to have been the line of the first extended tram subway, but logistical reasons dictated work began on line 1 instead.

Construction on this sector was especially challenging given the presence of the main sewer and numerous feeders. Line 3 was unique amongst the pre-metro sections in several respects. Its interface with existing tramways at either end was more complex, and at the south end problematic, as we shall see. The stations were on a more generous scale than on either of the earlier lines and incorporated different features. And it involved formidable constructional challenges.

At Gare du Nord the post-war arrangements for connection between tram and train at the new station were the converse of those at Gare du Midi. Here the tramway tracks deviated from the road, ascended a ramp, and occupied a spacious plaza at the same level as the station concourse. Interchange was thus exceptionally convenient and step free, although the lack of shelter made it an inhospitable place on wet and windy days. All this was swept away in 1974, ready for construction of a high-rise building immediately west of the station frontage, which it sadly obscured. This contained office space and a bus and coach station, providing a full inter-modal facility. In effect the new Gare du Nord pre-metro station occupied the basement of this building, and its construction necessitated extensive temporary works and diversions, which began on 6th August 1974. For the first time in recent years the terminating service 90 cars, which had formerly turned using tracks in neighbouring streets, reversed in a stub terminus, requiring double-ended trams of type 7700.

At Gare du Nord, as also at Rogier, the new underground layout brought into use on 4th October 1976, anticipated future developments and is an instance of the long-term planning which lay behind the whole pre-metro project. The station had provision for four platforms although not all were initially commissioned.

At the southern end of the new line 3 the arrangements were somewhat more tentative. The ramp leading from the 1957 subways to Boulevard Lemonnier was a temporary structure, and was taken out of use on 12th April 1974, after installation of extensive temporary trackage by-passing the construction site to the north. These diversionary lines ran from near the future Anneessens station via the parallel Rue d'Artois (southbound) and Rue de la Caserne (northbound), enabling cars to and from Midi to exit the subways via the Poincaré ramp. The north-south tunnels were then carefully excavated and the existing ramp altered so the new tracks could join end-on with the existing Lemonnier station; this took on a strangely diverse appearance, its southern end the original 1957 structure with foot crossings over the tracks, the new northern end more resembling a metro station. These works were finished on 4th October 1976 and the five services still using the tunnel could again access the Midi subways directly. This remains the arrangement in 2018.

It was, however, somewhat unsound, but was made necessary both by cost considerations and by the immense works which would have been needed to produce a more radical solution. The essence of the problem was the presence of the three flat junctions within the tunnel where the branch ramps to Jamar and Poincaré connected with the north-south tracks. This subway section was also still unsignalled. When the time came fifteen years later for the extension of line 3 southwards the problem was partially evaded again, and its rectification will be one of the most demanding tasks in the conversion of the line to full metro operation.

The four intermediate stations on the new line 3 (Rogier, De Brouckère, Bourse, Anneessens) were unique on the pre-metro. As well as normal side platforms, partly cut away to allow 'temporary' use by trams as on lines 1 and 2, each station also incorporated a central island platform. The intention was that in due course, under metro operation, these would facilitate lateral passenger flow, with travellers exiting on one side of the cars and boarding on the other. This arrangement was long delayed, but was eventually introduced for tram passengers in 2002-4, after part of the central high platforms had been cut away, and new access arrangements constructed from the mezzanine level. At De Brouckère and Rogier the line 3 stations were part of larger works which also catered for the intersecting lines 1 and 2 respectively.

Rogier was a particularly interesting example of the long-term planning considerations already mentioned. As well as platforms on intersecting lines 2 and 3, there were also terminal stubs at an upper level, served by a parallel pair of tracks from Gare du Nord, where there was a non-conflicting junction with the main line 3 tracks. The intention of this was

The post-war relocation of Gare du Nord a long walk away from its original site facing Place Rogier further justified investment in the North-South tunnel. This remarkable view in 1950 shows the old station under demolition in the centre, with Place Rogier in the far distance. The new terminus is behind the camera, and the street on the right-hand side is Rue du Progrés below which the tunnel was later excavated.

The new Nord station was served by trams on an elevated plaza just outside the concourse. The tall buildings on the left beyond the ramp occupy the site of the old Nord terminus, with the Martini Tower dominating the scene. [MJR]

Rebuilding the west front of Nord station to accommodate both the pre-metro tunnel and the superimposed Communications Centre was a long and delicate task, with need for tram deviations as seen here in May 1975. The 'Brussels Manhattan' is in the background! [MJR]

An informative view looking north on the site of the former Nord Station Plaza in April 1976, showing the temporary tracks, including a reversing siding (centre) for service 90, which therefore had to be worked with double-ended cars. The new pre-metro station is under construction far below on the left, and in the centre background are the remaining buildings in Rue du Progrés where the northern portal would be built. [MJR]

The core of the North-South tunnel was excavated below Boulevards Anspach and Lemonnier, with temporary trackage and road surfaces in place for several years. We are looking north here towards De Brouckère on 29 May 1972, with the Bourse to our right as a motor and trailer set head for Nord. [MJR]

We are in Rue du Progrés in September 1976, looking north, with the throat of Gare du Nord to the right and the excavations for the new tunnel and its north portal just behind us. A '4000-type' improvised four-axle articulated car from Vilvorde heads south (the red service number indicates an 'out-boundary' service with extra fare). [MJR]

Rogier pre-metro station –seen on 22 July 1979- was especially complex, with intersecting through tracks on two levels as well as this terminal section above them, intended for residual tram services from the north when the pre-metro was converted to metro (as is now expected). The bay on the far left was used to display historic rolling stock. [MJR]

Turning round, some years later, we are now looking back towards Gare du Nord with the Progrés portal in the background and the new reversing siding in the centre. A six-axle car heads towards Bordet on busy service 55.

A substantial deviation was required at the south end of the North-South tunnel, to allow reconstruction of the Lemonnier portal. A long avoiding route was built, leaving the existing surface lines near the present Anneessens station, with separate inbound and outbound lines in adjoining streets. A northbound PCC leaves the inbound track on 15 April 1974 passing the statue of Francois Anneessens himself (a patriot who was executed in 1719). [MJR]

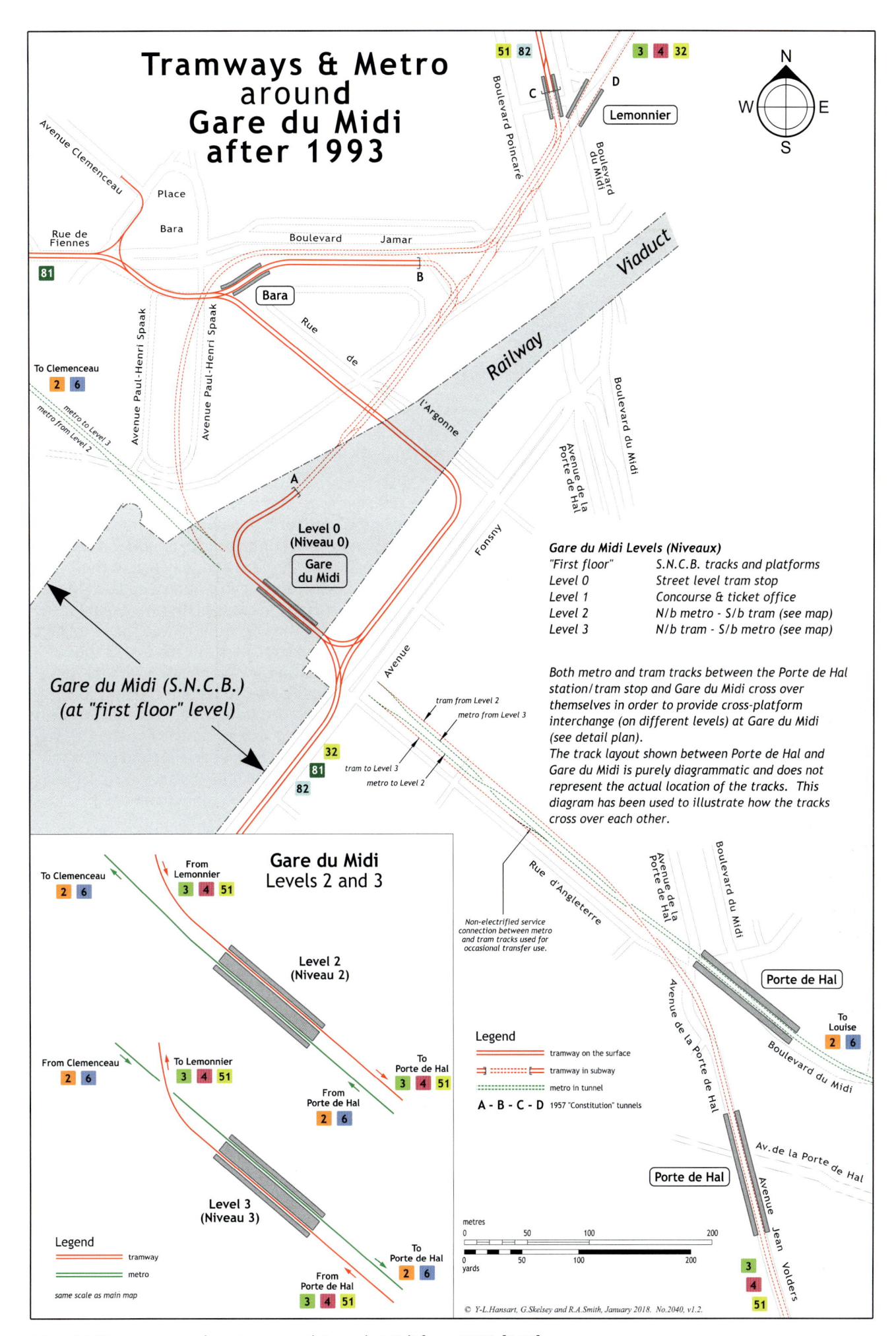

Map 15. Tramways and metro around Gare du Midi from 1993 [RAS]

to allow the residual tram services from the north of the city to terminate when line 3 was converted to full metro. Cross-platform interchange between trams and metro would be available at the four-track Gare du Nord station. With planning now proceeding for the conversion of the line this long-delayed intention will at last be realized.

These arrangements, which can puzzle observant travellers, will be apparent from map 34.

Excavation of the very large station chambers, notably at Rogier and De Brouckère where they occupied two levels plus a mezzanine, required long interruption to normal surface activity and damaging impact on local businesses. Those who remember, say, Place de Brouckère in the years 1972-76 will recall the temporary road surfaces and intermittent disturbance. Tram traffic had to continue at surface level along the artery, as there was then no possibility of assembling sufficient buses to replace trams. At the northern end temporary diversions were introduced at the end of August 1974 diverting trams to and from Rogier via Rue St-Michel and into Boulevard E. Jacqmain, allowing construction work to take place at the Adolphe Max end of Place de Brouckère.

The work was eventually completed and on 4th October 1976 all tramway traffic was diverted into the new tunnels. This was a defining moment for the city's tramways, removing the last evidence of street operation from the heart of the city, and the closure of the surface lines was marked by a procession of historic vehicles. Time for the 3.3km run between Nord and Midi was reduced from seventeen to nine minutes, and interchange was facilitated by a travellator between lines 1 and 3 at De Brouckère. Of course trams still ran below ground and along the Rue Royale lines further east, but were invisible to many, including the tourists thronging the Grand' Place. There are visitors who now believe Metro and pre-metro are the same, and that trams as such no longer run in the city.

At its northern end the new line rose towards the surface on the site of the former ramp, and joined the original subway. Here, rather incongruously, Lemonnier station now consisted of a quasi-metro station at its northern end, and the original 1957 tram stop to the south, including a foot crossing between the platforms. In the course of the complex and long-drawn-out construction here a building at the apex of Boulevard M. Lemonnier and Boulevard Poincaré was demolished and replaced by a new structure incorporating the station entrance on the ground floor. All this survives in 2018, although somewhat battered by time.

An extension southwards of line 3 had been envisaged from an early date, and a south-west/north-east cross-city line joining Uccle and Schaerbeek had been sketched in. In the interim a simpler but still significant project took effect after 1988. This extended line avoided congested streets, and in particular the spectacular roundabout at Barrière de St-Gilles.

This 1993 line was joined to the existing Constitution tunnel in the north and to new surface tracks in the south, opening up a continuous underground tram line between Gare du Nord and the edge of the southern suburbs. Starting in a new deep-level Midi station, the trams made cross-platform interchange on two levels with the line 2 metro, the tracks rolling over each other underground at each end of the station so that the tramway connected with trains running the opposite way, to benefit the principal direction of travel: northbound trams connected with eastbound trains, and vice versa. The line then served intermediate stations at Porte de Hal (also linked to the adjacent platforms on line 2), Parvis de St-Gilles, Horta, and Albert, where the tunnel branched.

The western branch came to the surface in Avenue Jupiter and passed over new street tracks in Avenue Bertrand and Avenue van Goidtsnoven

At Lemonnier the 1976 line rose into the 1957 station, where there was a strange juxtaposition of pre-metro and tramway, including this foot crossing between the platforms.

In connection with the 1976 construction of line 3 a new station entrance was constructed for the altered Lemonnier station. This remains, somewhat battered, today. The 1957 ramp towards Boulevard Poincaré is in the left background. Note that the 'M' for Metro identifiers are used for pre-metro stations also.

before rejoining the existing Chaussée d'Alsemberg tramway towards Uccle and (originally) Silence. The line restored tram service to the locality known as 'Altitude 100', and tracks were replaced in Avenues Jupiter and Bertrand from which they had been removed in 1961. The eastern branch from Albert rose to the surface in Avenue Albert and joined existing centre-reserved track leading towards Vanderkindere and the Avenue Winston Churchill line, a high-quality tramway.

The 1993 stations again showed the characteristic features of the earlier pre-metro stations, with lowered sections to allow easy boarding to trams. They displayed notable artistic features, in particular at Horta which commemorated the notable Belgian art-nouveau architect Victor Horta and incorporated metal work from one of his principal works, 'La Maison du Peuple'. Albert station, partly in the apex of the junction, has a slightly awkward interim arrangement, but incorporates provision for a future Metro terminus and for connecting tramways as already existing at Simonis.

The line was formally opened by HRH Prince Philippe (now King of the Belgians) on 3rd December 1993, adding another 3km and four stations to the underground network. The remaining surface tramways south from Gare du Midi, along Avenue de la Porte de Hal and Chaussée de Waterloo were abandoned

As mentioned these works left partly unaltered the original 1957 Constitution tunnels at Gare du Midi, which continue to serve as a slightly incongruous link between the surface tramways, the north-south tunnel, and the vast splendours of the two-level deep level station below Gare du Midi. Although the 1976 and 1993 tunnels and stations can readily be converted for Metro use, following the well-practiced techniques of earlier change-overs, the Constitution tunnels will have to be by-passed.

5.44 Line 5 (Eastern Orbital)

The Outer Ring tramways around the eastern side of the city have developed as one of the finest layouts on the system, mostly with segregated track and offering a fast and reliable service. The pattern of use has evolved over time, the lines now forming less part of a mainly radial system but instead offering orbital inter-district travel. This has been influenced partly by the growth of office premises along the route (for instance, around Square Montgomery) and also the flow of students to and from the Université Libre de Bruxelles (ULB) south of Buyl. At an early stage in metro planning there was an intention to build a pre-metro, later metro, initially along the larger part of the north-eastern quadrant. This was designated 'line 5'. Although there had been considerable investment in improved tram tracks the problem continued of interruption at major intersections, such as those at Princesse Elisabeth, Place Meiser, Square Montgomery, Chaussée de Wavre, and Buyl. Montgomery in particular was a complex junction, and construction of short local underpasses there was included in the first pre-metro plan. Work was to be associated with complementary road tunnels.

In the event a more radical solution was adopted, and a 1.08km tunnel was opened in 1972 forming an underpass and station at Diamant, avoiding the cross-traffic in Avenue de Roodebeek and the junction with the E40 motorway. This was extended southwards in 1975, with a further 1.5km of tunnel and three additional underground stations at Georges Henri, Montgomery, and Boileau. At the last-named a street interchange was possible with Thieffry station on the Herrmann-Debroux branch of line 1. From 1976 Montgomery became one of the most complex nodal points on the rail system, with line 5 in a north-south direction, the new line 1 metro extension running east-west, and a remaining surface tramway on the surface, with a new turning loop for what is now

In January 1958 the tramway works for the Exhibition were almost complete, including the new STIB and SNCV tracks through what is now De Wand station, and the Vicinal tunnel beyond. An outbound Vicinal set ascends the exit ramp towards Grimbergen.

Meiser was a weak spot along the line 5 pre-metro as its busy junction lay just north of the portal requiring cars to cross busy roads on the level. This has not yet been rectified. [MJR

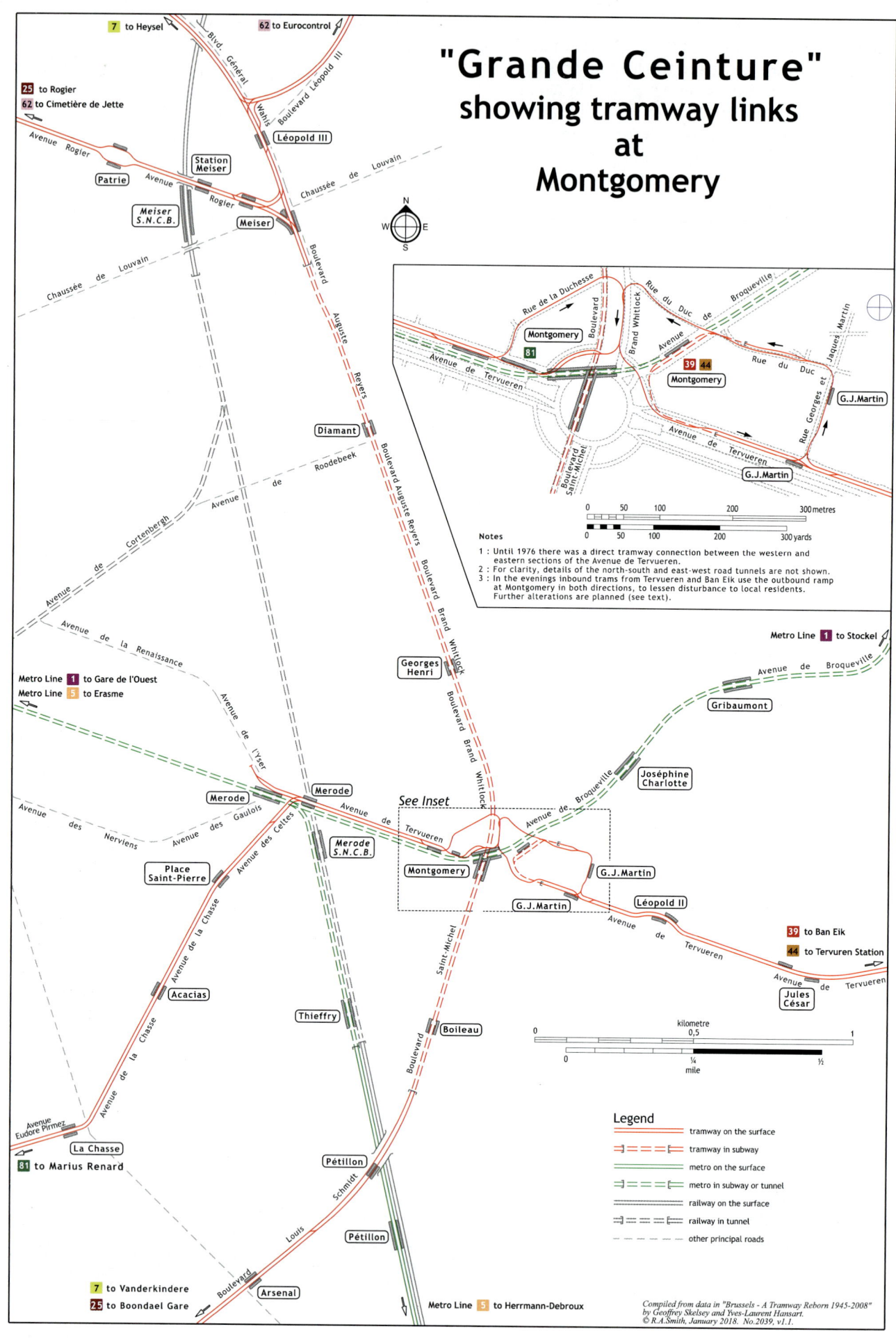

Map 16. Tramways and metro around Montgomery: 'grande ceinture' [RAS]

In the decades since 1958 trees have grown around the tracks leading to De Wand and the short underpass below Avenue des Croix de Feu.

service 81 in back streets. The feeder tram services to and from the east (39 Tervuren, and 44 Stockel) were cut back in 1976 to terminate in an underground loop (see map 16).

The worst remaining feature of the orbital tramway is the layout at Place Meiser, where the tunnel terminates with a temporary ramp just south of the square, leaving trams to cross the intersection on the level. The stopping point in the apex of the junction is awkward and congested.

5.45 Minor Underpasses

For completeness we need to mention two minor underpasses and a longer tunnel, all of which originated in works associated with the 1958 Exhibition at Heysel. The thorough improvement to the long STIB line beside Avenue des Croix du Feu on the eastern edge of the site, built for the 1935 Exhibition, included underpasses at Pont van Praet and Gros Tilleul opened in 1957. The latter included a new four platform low level station (now called 'De Wand') interchanging with the Vicinal lines serving Grimbergen and beyond. The sharply-curved tunnel at Pont van Praet made thrifty use of a long-disused former railway tunnel built for a private branch line into the Royal park at Laeken Palace.

Nearby the relocation of the Vicinal lines to the south of the Exhibition area included a major new tunnel taking the line below the edge of the site close to the vast French pavilion. This 520m structure included a subway station with stair access serving one of the Exhibition entrances. Disused after the abandonment of the last Vicinal services around Brussels in 1978

When the Esplanade line was upgraded for the 1958 Exhibition the opportunity was taken to utilise a disused railway tunnel to carry the tracks below the intersection with Avenue Van Praet.

the tunnel stood idle until, in an inspired move, it was incorporated into a new and popular STIB peripheral line (service 19) stretching around the north-western quadrant of the city and acting as a Metro feeder in several places. This was brought into use in 1994, together with reordered and improved approach tracks to the south and a new stub-terminus in the former Vicinal area of De Wand station. There have been several proposals for extension of STIB services over the former metre-gauge route towards Strombeek and Grimbergen.

For over fifty years there have also been plans for further underpasses to avoid notorious pinch-points elsewhere on the system. In their most recent iteration these included short tunnels at Meiser and Le Grand along the eastern peripheral route, at Louise to avoid the constricted street at the inner end of Avenue Louise, and at Place Liedts. These have so far foundered on the twin difficulties of huge cost and the problem of maintaining traffic during a lengthy and disruptive construction period.

5.5 Pre-metro Operation and Rolling Stock

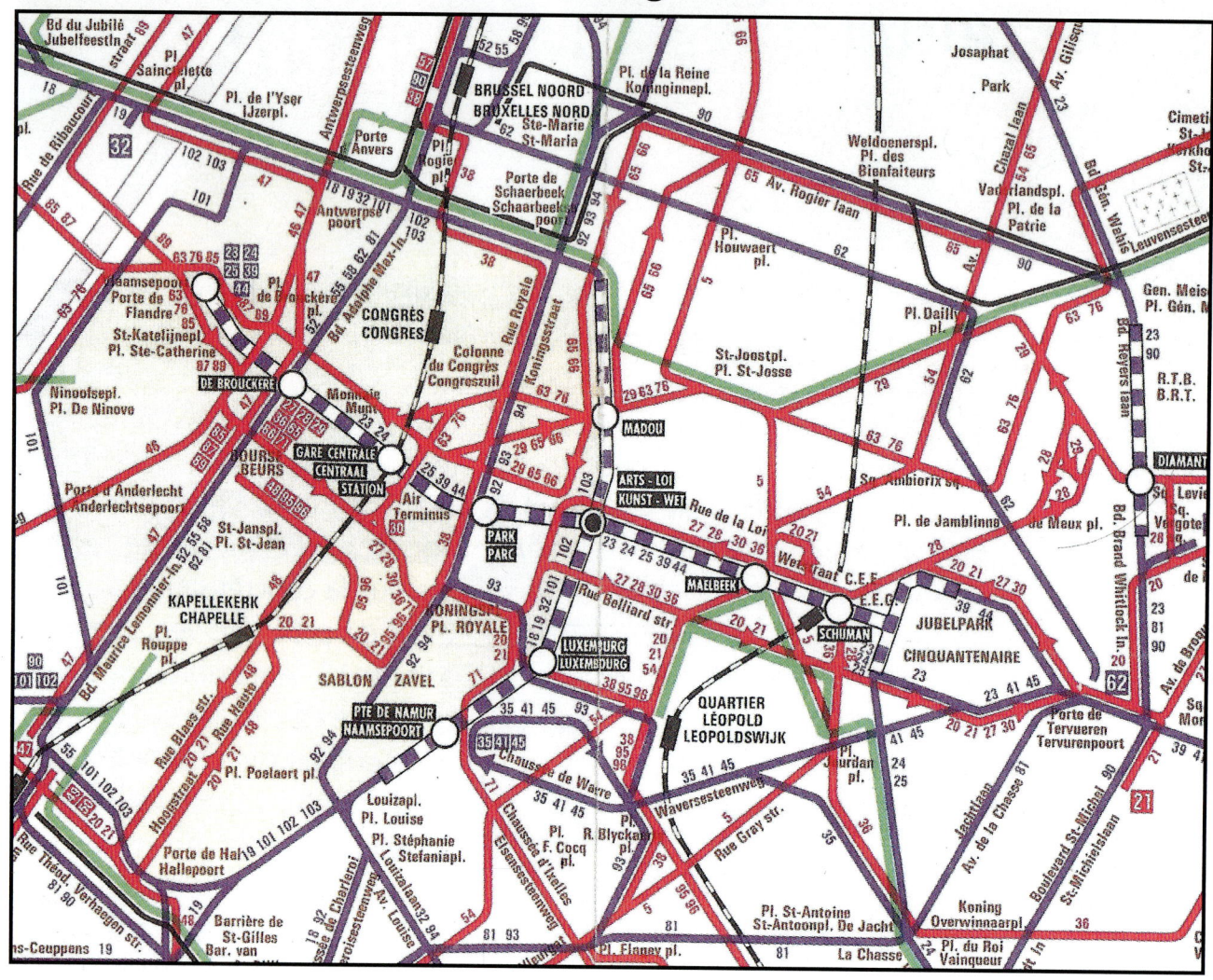

Map 17. An extract from the STIB's 1972 route plan shows the full pre-metro network as it then was, before the first conversion (of line 1 to Metro in 1976).

5.51 Operation

The inauguration of tram working through the new tunnels in 1969 required radical change in many aspects of tramway operation as it had evolved over sixty years. Detailed new regulations were required, covering such topics as signalling and power failure, aspects of signals, station routines, and emergencies. Further training and certification was needed, as were new techniques for track and overhead maintenance. The radio communications system enabled constant contact with the control centre (a novelty at the time for tram drivers) and there were back-up arrangements. The controllers could intervene to bring trams to a stand in general emergency. Battery lighting was available at stations and in tunnels, with automatic change-over if needed. Extensive fire precautions were installed.

Essential to the success of the venture was the transfer of fare payment from cars to stations, both to reduce fraud and to limit dwell-time at stations and hence to maintain a consistent service. In practice regularity was not always good: slow boarding and alighting, especially in crush loading conditions, caused bunching and a theoretical 40-second headway was often exceeded, with cars stacking on the approach to busy stations.

Traffic delays arising beyond the two eastern portals were translated into the tunnel operation as there were then few priorities for public transport. On Line 2, in addition to underground stations, a pre-payment system was introduced at an enclosed surface station at Louise, to reduce dwell-times at the busy junction.

In the subways cars normally ran to the extreme end of the lowered portion of platforms –to the puzzlement of casual passengers who waited in the middle. The signalling system allowed a second car to enter a platform slowly under caution. Dwell-times could be further lengthened as passengers hurried to board their chosen car. Mirrors were fitted to enable drivers to observe the doors.

Overlooking the turning circle and sidings at the temporary Ste-Catherine terminus was an elevated signal cabin of rather traditional appearance, which controlled operations there as well as being a supervisory centre to deal with staffing and other issues at a busy location.

Typical of pre-metro operation is this view of Georges Henri station on line 5, showing the platform arrangement, signalling, and train control equipment. [MJR]

5.52 Rolling Stock since 1957

When the Midi subways were opened in 1957 much of the tramcar fleet was made up of two-axled 'standard' motor-and-trailer sets built before the Second World War, most of which could work through the tunnels. Starting in 1946 many of them had been refurbished and modified, including addition of automatic doors and internal reorganization with a seated conductor and greater standing space. The only bogie cars before introduction of the 7000-type PCCs were the 25 cars built in 1935 for the Heysel Exhibition (type 5000). In 1957 there were about 600 motor cars and 300 trailers.

No new cars were ever specifically built for tunnel operation, but pre-metro operation was restricted to cars with automatic braking equipment. In this section we describe the car types in operation down to the present day.

The following table lists the STIB fleet as it was during the initial pre-metro era.

Year	2 axle trams	Bogie trams	Articu-lated trams	*Total motor trams*	Buses	Metro units
1970	336	197	44	*577*	491	0
1976	101	172	171	*444*	571	45
1980	43	172	230	*445*	607	57

Source: *STIB Rapports Annuels*

Two small groups of the two-axle standard cars were modified in 1967-8, with minor technical improvements such as built-in roller-blind front destination signs in place of wooden roof-mounted 'plaques'. Fifteen of the cars (1501-15) were equipped for one- or two-man operation as required, with appropriate alterations to the control equipment. Nine trailers of type 600 were also refurbished in 1968 to run with the 1600-type rebuilt motors. It had been intended to refurbish 55 motors and 40 trailers in this style, with a view to using them in the pre-metro subways, but this idea was not proceeded with. None of the two axle cars or their derivatives worked through the pre-metro tunnels, where wooden-built cars were understandably banned (as they were from, for example, the Kingsway tunnel in London).

There were also attempts to modernise the interiors and appearance of the elderly fleet whilst awaiting new trams, and these cars worked through the Midi subway but not the pre-metro's. After an initial trial with a composite bodied conversion, 100 cars were rebuilt in 1959 – 61 from standard trams 4001-4100 originally constructed between 1929 and 1933, renumbered in the '**9000**' series. They received new metal bodies, resembling an abbreviated 7000-type. Modern seating, fluorescent lighting, and electro-pneumatic powered doors were fitted. Originally a conductor's position was fitted in front of the rear doors, but this was removed on conversion to one-person operation from 1964. The cars had a pleasant and up-to-date appearance but obviously lacked the technical refinements of the modern fleet. They did not haul trailers (none of which was modernised in this form) and were withdrawn by 1980.

Forty-three four axle single-ended articulated cars were built in 1963-6, converted from pairs of standard two-axle trams with an added suspended centre section, and numbered in the '**4000**' series. The long gestation period of new PCC-based articulated trams (later to be types 7500 and 7700) was operationally inconvenient and this reconstruction project provided a relatively low-cost articulated fleet in the interim. An important purpose was the replacement of trailer operation on busier services, with consequent saving of staff. The new bodies again outwardly resembled the type 7000 PCCs, with three pairs of double doors and two-and-one seating in the front and rear sections. The cars were fitted with a conductor's position in front of the rear door, but they were equipped for one- or two-person operation as required. Withdrawal began in 1979.

The 172 '**7000**' class, four-axle single-ended PCC cars, delivered between 1951 and 1971, this series

We take a glance at typical rolling stock used on the system in recent years, beginning with the ubiquitous 7000-type PCCs. This car is running inbound in 1963 along Bvd du Jubilé and will shortly join Bvd Leopold II towards Rogier. The tracks are still in use today, but since 1988 no longer connected to the Ceinture lines. [Courtesy John Bromley]

The first of the six-axle articulated cars were originally single-ended, based on this prototype seen approaching Rogier from the west in 1967. [Courtesy John Bromley]

The big eight-axle cars in type 7900 are still valuable for their capacity and reliability. 7954 is seen in 1987 near Hotel des Monnaies on a section partly replaced by metro in 1988, and finally abandoned in 1993.

was derived from the classic American 'Presidents Conference Committee' trams developed in the 1930s. The body design was different however, narrower and with tapered ends, but technically the cars initially followed American practice. Similar cars ran in St-Etienne, Marseille, Gent, and Beograd, and still run in Antwerpen. During the planning stage for the pre-metro consideration was given to installing multiple-unit equipment, which would have enabled coupled sets to provide greater capacity in the tunnels: this was, of course, implemented with similar cars in Antwerpen. In the event no action was taken, perhaps because of objections to coupled sets running in the streets. The cars were the mainstay of the first pre-metro section. (These and the two succeeding types are fully described in our book PCC Cars of Western Europe (2011). In the initial stages of pre-metro operation cars equipped with automatic braking equipment were identified by a 'T' symbol below the headlight.

The 127 '**7701**' class six axle double-ended PCC-based cars. This fleet was derived from an experimental 'Eurotram PCC' developed by BN in 1962 (later nicknamed 'Caroline') and applying PCC technology (trucks and electrical equipment) to a single-articulated body. The cars originally fell into two broad groups, 98 single-ended cars delivered in 1971-2, numbered 7501-7598; and 30 double-ended cars (7801 to 7830) delivered in 1972-3. The single-ended cars were built with provision for conversion to double-ended format, and they were so rebuilt in 1981-86.

The 7000-type trams had incorporated American-style PCC bogies, either built new or purchased second-hand. For the production articulated cars the builder BN introduced a new bogie incorporating features of the PCC B2 and B3 designs, but with an upper frame resting on rubber journal springs to improve vibration damping. The central portion of the frame supports the bolster through coil springs and rubber. These bolster springs are canted slightly outwards to control lateral movement. These innovations produced a superb ride quality and quiet operation, aided by the use of Swedish-designed resilient wheels.

Design of the production articulated fleet had been delayed by the need to devise a PCC-based control system capable of controlling six traction motors. This was eventually achieved by the use of two separate accelerator units, one in each section of the tram, making it possible to introduce an articulated car with all axles motored, rightly regarded as essential for Brussels' operation.

Delivery of low-floor cars is bringing to an end the long reign of the six-axle cars, and at the end of 2017 only 65 remained in service.

The 61 '**7900**' class eight-axle double-ended PCC-based cars were delivered in 1977-8 as a further development of the six-axle articulated fleet. They are notable for their large windows and distinctive seating. These very long cars are fitted with two pantographs and are effectively two PCC cars back to

As described in the text the 2000-type low-floor cars were not unproblematic but they brought new standards to the system. 2011 is seen in Rue de Stalle in 1994. [Courtesy John Bromley]

back, operating in double traction. Proposals were made in 1978 for a new fleet of 235 stadtbahn-type eight-axle hybrid cars numbered in the **10,000** series for service on both high- and low- platform sections. They are described more fully in section 6.3, but the initial order was cancelled.

The fleet of 51 low floor trams in series '**2000**' entered service in 1993-95, after lengthy experiments and trials; this was the first new class of trams since 1977-8 and they were a striking indication of the transformation of the tramway system. They were assembled at the BN Brugge plant, incorporating components from other sources. This unique Belgian design of six-axle, articulated, double-ended low floor cars was introduced to rehabilitate the improved north-south tram routes serving Rue Royale. They are derived from experimental work carried out by BN for a new intermediate capacity transit vehicle to compete with the French 'VAL' system, known as TAU ('Transport Automatisé Urbain') and proposed for service in Liege. An articulation bogie ('BAS 2000') was designed for it, the basis for type 2000's innovative bogies.

The outermost bogies of the units have two large 640mm motorised wheels and two smaller 375mm carrying wheels (in appearance rather like the old 'maximum traction' bogie), and the centre bogie has four motorised wheels, all motors being mounted on the wheel hubs. Motors are three-phase, giving the cars a distinctive sound when accelerating and braking. The bogies have no axles, enabling the car floor to be lowered between the wheels, although accommodating the equipment means that the outermost seats in the tram have to be raised on awkward 'podiums'. These trams reintroduced hand control to Brussels in place of the pedal control common to all the PCC-based cars.

The type 2000s are wider than standard Brussels cars (230cm compared with 220cm) and are also somewhat heavier than earlier six-axle trams. Until 2006-7 they only operated on certain selected routes, excluding the tunnels as they were not fitted with automatic signalling equipment.

> *In September 2003 tenders were accepted from Bombardier for a new fleet of 46 low-floor cars, to supplement the fleet and to replace the remaining 7000s.*

Perhaps the most striking change introduced with car 3001, delivered late in 2005, was a new livery, predominantly grey and brown, which began to supplant the long-established yellow colours, introduced by the Tramways Bruxellois although modified in tint and layout. The new livery was also applied to older trams, but not to the older Metro trains.

The order comprised –initially-- 46 double-ended, fully low-floor, modular cars (of the Flexity 'Outlook' model previously branded 'City Runner', first introduced in Linz). The trams were largely built at Bombardier's Brugge works, but using bogies and motors imported from other plants. 19 of these cars (type **4000**) were 43m long, and 27 shorter, at 32m (type **3000**). 22 further cars in class 3000 were subsequently ordered, making 49 of this type in all, of which about 30 were in service by the start of 2008. In January 2008 the STIB announced a third order for an additional 87 3000-type (32m) trams, for delivery in 2009-12. In 2016 there were 150 type 3000 cars and 70 type 4000, with further orders to come.

The 'Outlook' design uses 560mm diameter wheels with conventional axles driven by small electric motors mounted on the truck frames. The low-floor entrance sections are 365mm above rail level and the car floor slopes gradually up by 100 mm to the slightly higher sections over the trucks. The car is thus wholly multi-accessible. The trucks are mounted directly on the car body and this can result in poorer riding conditions: arguably no newer trams have matched the original PCCs for journey quality, and there are adverse effects on track condition. The interior layout of these trams is unusual, combining longitudinal opposite transverse seating in part. The finish is of exceptionally high quality, with leather upholstery and careful attention to décor and lighting. A fifteen-year contract with Bombardier includes overhaul of the trams' bogies and propulsion equipment.

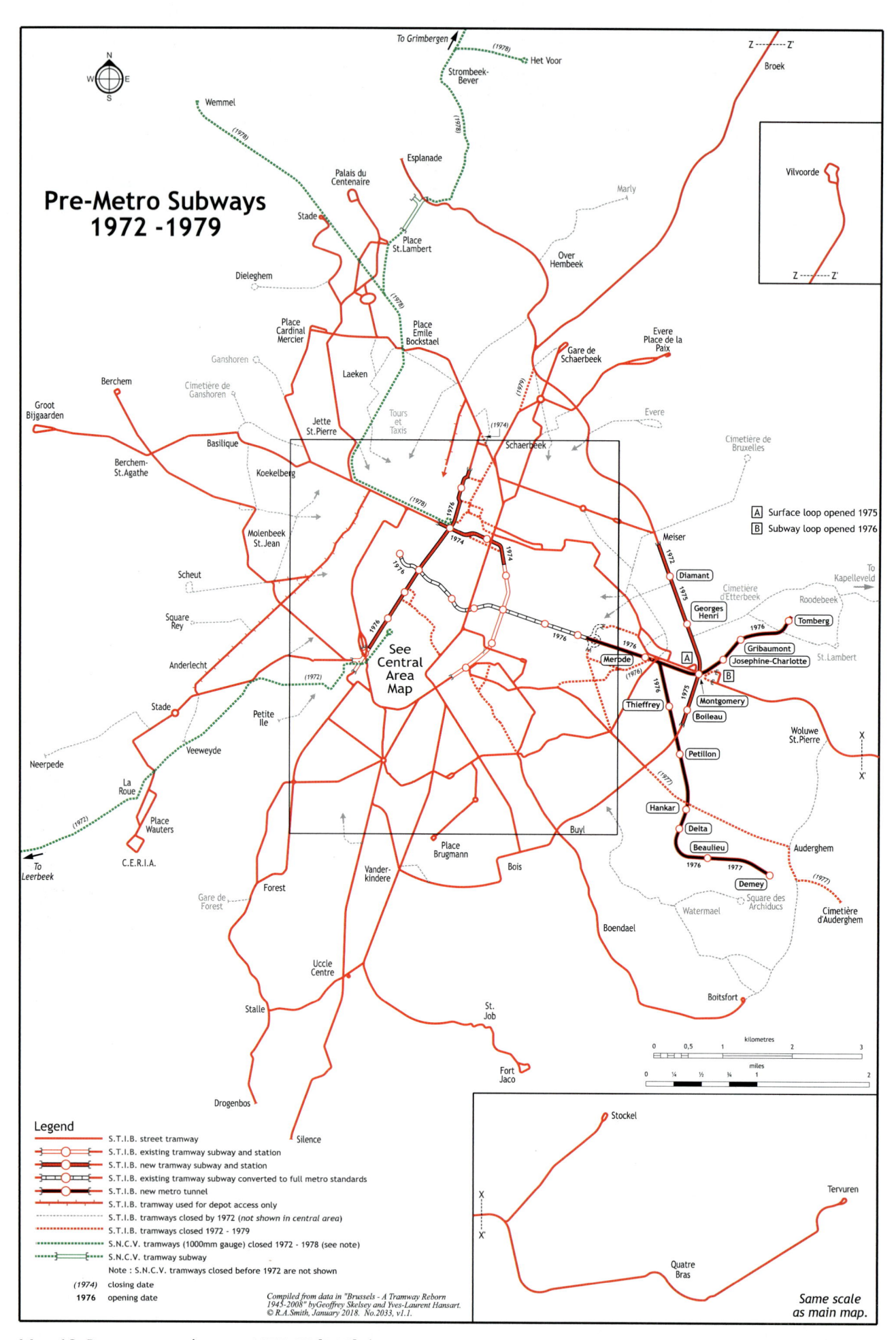

Map 18. Pre-metro and metro 1972-79 [RAS] the system.

6. Conversion to Metro from 1976

When the first Brussels pre-metro plans were aired in 1963-4 the duration of the initial phase, before conversion to full Metro, was left vague, but given the scale of interim works, such as the elaborate ramps east of Schuman, must have been assumed to have been fairly long. As early as 1966-7 this expectation was already changing, with partial Metro operation then expected by 1975. There were several reasons for this, most persuasively traffic analysis which showed that the flow on the east-west line, at least, would exceed 14,000 passengers per hour per direction, beyond the ability of single unit trams on a 40-second headway to handle. We must now turn to the long-drawn-out process of conversion from pre-metro to full metro, which long ago took effect on lines 1 and 2 and is now in prospect for line 3.

Conversion of line 1 from pre-metro also involved construction of extensions diverging at Merode. This is the scene during work at the site of Delta in 1975, with the SNCB east-side line, alongside which the metro would run.

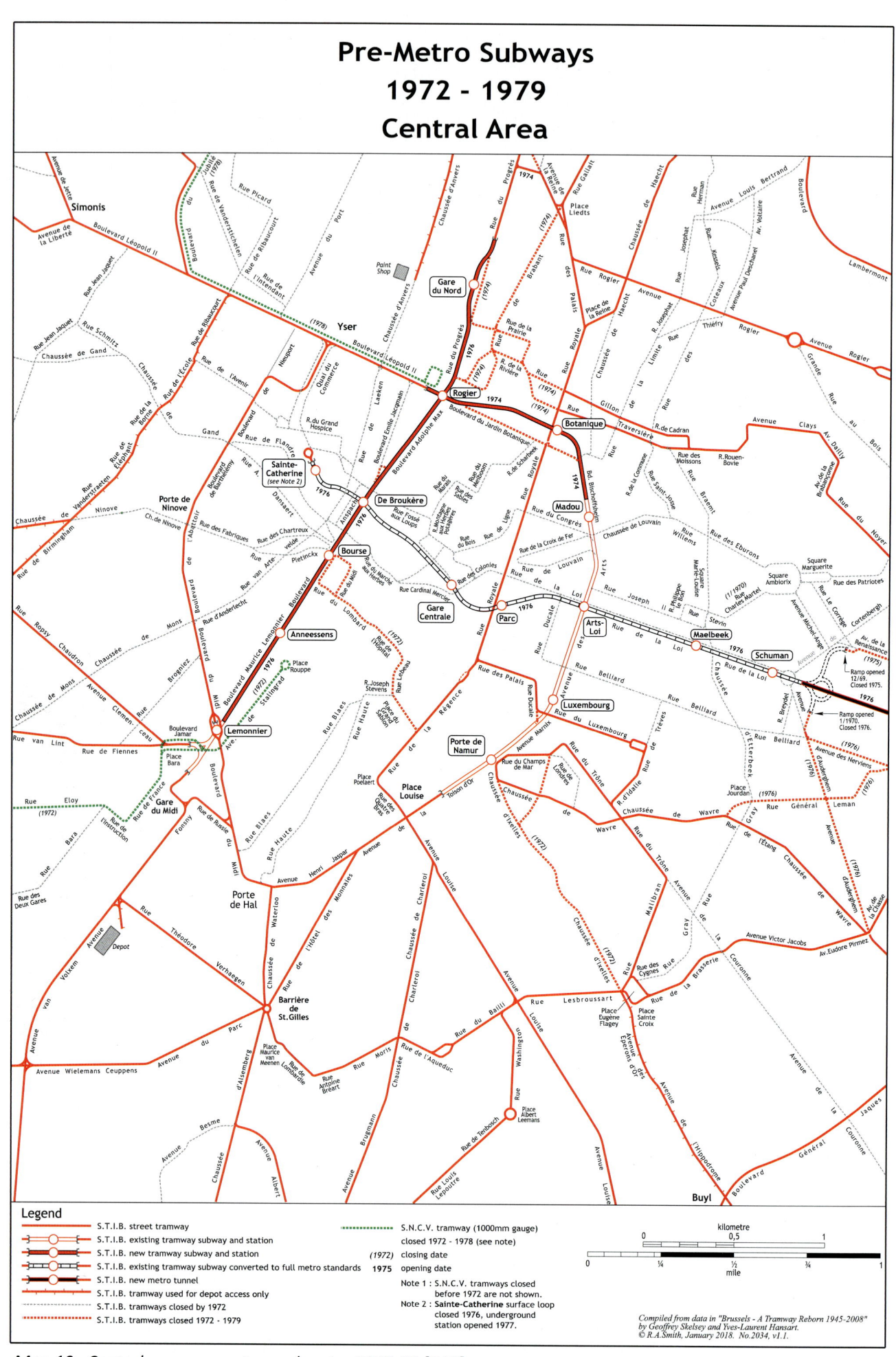

Map 19. Central area pre-metro and metro 1972-79 [RAS]

6.1 Line 1

The 1963-4 plans envisaged a steady extension of the tunnels on the east-west axis, with extension beyond Schuman and Montgomery, to link with existing high-quality tram lines on the Tervuren and Stockel routes. At the western end there would have been two tunnelled branches beyond Ste-Catherine, to connect with line 2 at Place d'Yser and extending beyond the canal to join the Chaussée de Ninove. The change of official mind arose primarily from what had been in the mid-sixties unforeseen developments around Schuman, arising from the rapid growth in EC (later EU) institutions which have turned what was a largely residential area into an extensive high-rise business district, with heavy commuting flows. The resulting growth in traffic from both directions rendered invalid the assumptions about the capacity of tram operation as it was then practiced. There was also perhaps a more generalised view that a city assuming such international importance required a Metro for prestige reasons.

The last tram through the east-west tunnels ran on Thursday 16th September 1976. Frantic but carefully-phased work followed, involving commissioning new signalling and power supply, disconnecting the remaining tramway ramp east of Schuman, and testing new rolling stock. Not all the lowered platform sections could be raised in time, and for a week trains ran with some doors disabled. HM King Baudouin formally opened the Metro on Monday 20th September, amidst considerable celebrations.

A special train ran from De Brouckère to Hankar, then returned to Merode before reversing through Montgomery to Tomberg. The events ended with speeches at the Monnaie Opera. The line opened to the public at 14h00, and was soon overwhelmed with traffic (travel was free). Only a single stub-end terminus was initially available beyond De Brouckère, which limited frequency to fifteen trains an hour. Within months traffic had increased by about 45 per cent above the pre-metro level, facilitated partly by more generous transfer arrangements covering travel from and to feeder tram and bus services

We will look later at the present form of the metro system (chapter 8), but to begin with the 1969 tunnels were extended east of Schuman, with a huge excavation at Merode to accommodate superimposition of the two tracks, thus facilitating a non-conflicting junction of the two branches: this required a temporary road viaduct above the works. The northerly branch initially terminated at Tomberg. This line was wholly in tunnel through a built-up district. At Montgomery there was new provision for trams still serving Tervuren and Stockel, cut back to serve as feeders. (A weakness of the arrangements here is that passengers connecting to and from the tram lines can only take every second train as the branches diverge one station west: in principle this could be addressed by re-extending the feeder tramway towards Schuman, thus offering again

Converting the line at Schuman in 1975 required closure of the northern of the two ramps east of the station, and cars on services 39 and 44 were diverted for the last year via the southern ramp to run along the south side of the Cinquantenaire park, as seen here at the divergence of the line to Porte de Namur. These tracks are now a bus road. [MJR]

To provide the best interchange at Montgomery between Metro and feeder trams an elaborate loop and underground station were built, including this inbound line and portal contrived in a corner of a residential area. This was originally needed because the services used single-ended trams, and now that they have all gone application has been made to create a simpler stub terminus using the present outbound ramp.

An evening Stockel car waits in the exchange platform at Montgomery: a ramp leads directly to the Metro platforms, and signals warn the tram driver of an impending arrival. [MJR]

one-seat service to the European district, but this is an elaborate and extensive project which has been planned but not yet programmed).

To facilitate interchange with the trams to Stockel and Tervuren (services 39 and 44 respectively), a short new terminal subway was built just east of Square Montgomery. Trams from Tervuren and Stockel approached this by new single line through residential streets with an entry ramp in a cleared area. Exit was by an ascending single-track ramp leading to Avenue de Tervuren. The single-line terminal loop was joined by an inclined pedestrian ramp to the inbound metro platform, and a signalling system was installed to warn tram drivers of approaching outbound trains. These measures went some way towards compensating for the loss of one-seat service from the prosperous eastern suburbs, and are an instance of a well-designed inter-modal interchange.

The southerly branch initially extended to Beaulieu, with a large rolling stock depot and maintenance plant at Delta (a site shared with the bus undertaking). Much of this line was above ground, occupying space alongside the state railway's east side peripheral line. 45 two-car metro units were on hand when the line opened (they are described later). At the end of 1976 the Metro comprised 9.8 route/km, with 16 stations.

In the west work was required to clear away the short-lived tram terminal at Ste-Catherine, where a 500m extension and a new underground terminus were opened on 13th April 1977. In the interim trains turned back at De Brouckère station. Construction proceeded to extend the tunnels to Beekkant.

Immediately after conversion an increase in traffic of 11.7 per cent was recorded, reflecting faster and more reliable journeys, as well as an end to delays caused by payment of fares on board. Traffic across the system also grew for the first time since the 1940s.

6.2 Line 2

The gradual development of pre-metro services along line 2 more exactly followed the original principles, and as we have seen successive underground extensions with temporary ramps took place over more than a decade. It was at first envisaged that the very difficult and slow tunnelling west of Rogier to Simonis (where there was a temporary ramp into an underground station) would be completed for tram operation, as would the equally difficult works south of Louise towards Gare du Midi. A late attempt to establish a comprehensive subway/surface pre-metro network came in 1978 when the STIB published fresh proposals. Looking at map 20, the northern part of the planned network includes both lines designed to avoid congested areas, and extensions to important new objectives. Note the by-pass line along Quai des Usines (completed in 2006) and an alternative route to Cimetière de Jette (which has not been built).

South of Louise on line 2 trams continued to run on roadside reserved tracks until 1985 when a new portal was opened near this point. In 1984 a 7000-type runs downhill from Louise in the distance, passing the construction works on the left.

After 1985 trams still used surface tracks past the ancient (but much renovated) Porte de Hal, the last of the city gates. On the left note the booth occupied by a 'controleur' or inspector, supervising movements at a busy location (and answering passengers' inquiries). Below ground work proceeded for several years on constructing the new tunnel to Midi. [MJR]

At the western end of the Ceinture routes an underground station at Simonis was completed in 1986, with two ramps leading in from the Basilique direction at its western end, and another leading to surface tracks towards Rogier. Here a car leaves the subway to the west on service 19 which then ran through from the Ceinture. After 1988 this became a feeder service, but this ramp is still in use. [MJR]

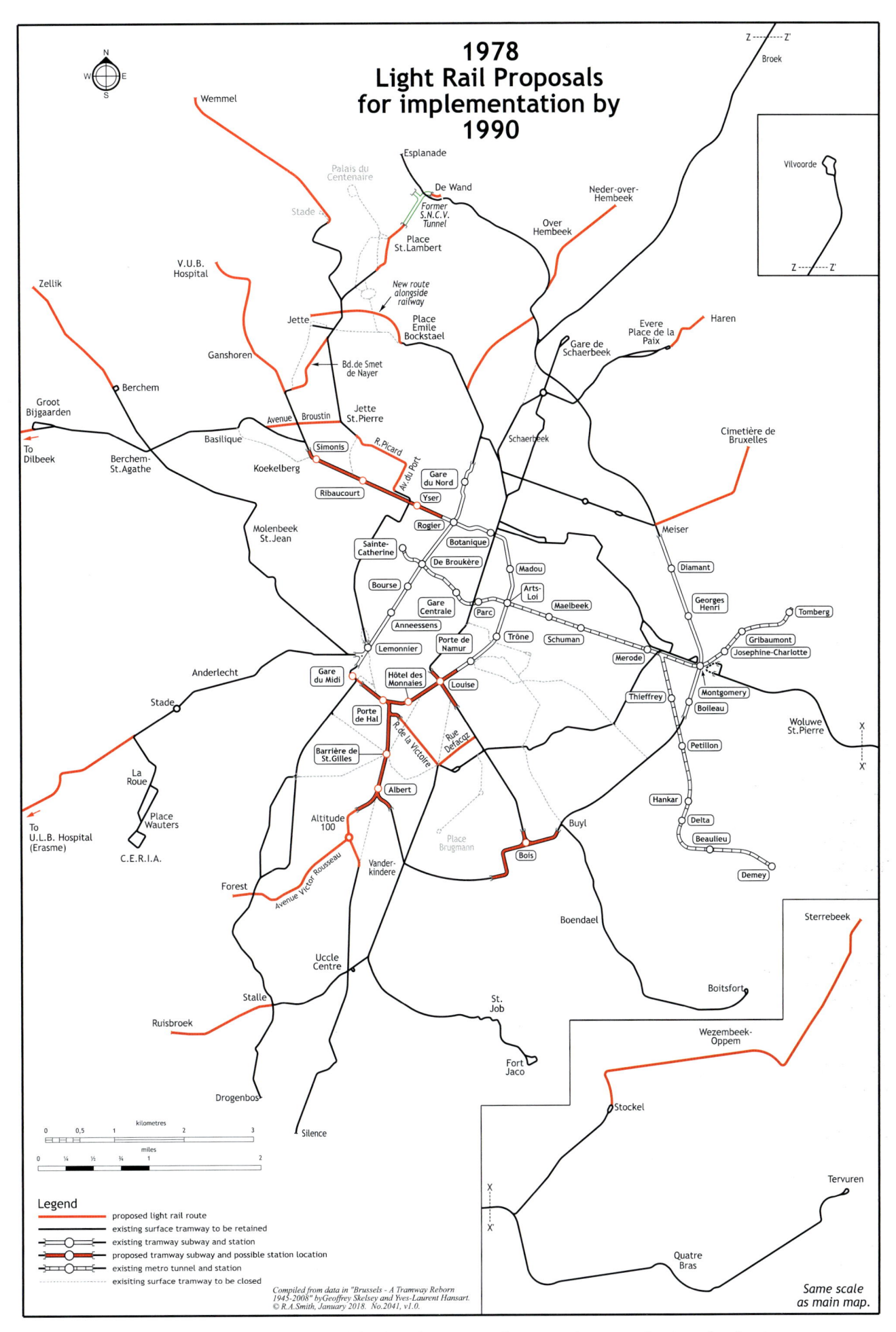

Map 20. One of the outline concepts published in the 1970s, this map shows aspirations for further development of a comprehensive light rail and pre-metro network, including feeders to line 2. Its conversion to metro in 1988 brought these ambitions to an end. [RAS]

Extensions include those through the former Vicinal tunnel at Heysel (opened in 1994), the branch to the VUB hospital at Jette (nearing completion in 2017), and along three former Vicinal lines (which have not yet been restored). These and other routes would have left line 2 at Simonis, maintaining a wide range of through workings to and from the suburbs.

South of the city there are also several ingenious new routes aimed at avoiding congestion in Chaussée d'Alsemberg and Chaussée de Charleroi, which never occurred. Especially notable is the extension to Erasme Hospital, which was opened in 1982 as a perfect example of a pre-metro surface line, with heavy duty track and catenary overhead.

All these lines would have branched off either the Ceinture pre-metro line, still to be worked by trams, or line 3, as extended south in 1993. They would have created a wider system of effective rail routes, exactly fulfilling the original specification. However, they would have required a new hybrid fleet fully to exploit the underground sections. As it was a new tram service '2' (so designated because this was pre-metro line 2) was inaugurated between Rogier and Midi on 24th November 1981, to provide much-needed additional capacity on the pre-metro sector of the Ceinture.

It was at first served by 7000-type cars sourced from Avenue du Roi, Ixelles, and Molenbeek depots: they were not really suitable for such heavy traffic and were sometimes commandeered when required by their home depots, which did not add to the reputation of trams on the route. Service 2 at first terminated at its northern end on a temporary surface loop west of Rogier, at Jacqmain: this was abandoned in 1987 when its site was required for construction of a new office tower, and thereafter double-ended articulated cars were necessarily used.

National and international events now impacted on the STIB's plans, and the era of assured and consistent public investment in Belgium (and elsewhere in Europe) was drawing unexpectedly to a close. The oil crisis which began in 1973, associated with cost inflation, decline in traditional industries, unemployment, and austerity, was to bite, and the formation in Belgium of a centre-right government under Wilfried Martens in April 1979 ushered in a more hostile time. The construction chronologies will show the slower pace which followed.

The initial order for the 10,000 type cars was cancelled (see 6.3), and in 1986 the unitary National Government (then responsible for national infrastructure works) decided to eliminate the transitional stage on line 2 extensions and to proceed directly to full metro operation. Several of the once-planned extensions have been built but do not form part of a complete pre-metro system.

The 1986 decision was overtly political. Transport Minister Herman De Croo was an able and ambitious politician in the second centre-right Belgian government formed in 1981. Although it differed in important respects from the policies of the contemporary Thatcher governments in Great Britain, the then Christian People's Party shared with them, ostensibly at least, fiscal conservatism and a rigorous attitude to public debt and deficit. Economy in executing public infrastructure works was therefore attractive to them and their supporters.

The problem from the point of view of those wishing to see tram operation continue on Line 2 was the need for new tramway rolling stock: there were at first insufficient articulated cars to provide the whole service and the 7000-types were unsuitable. A project for new hybrid rolling stock has been drawn up (see 6.3 below) but this would have required what was then central government funding. De Croo was able

The much improved Bvd Leopold II retained surface tracks for a short distance between Bvd du Jubilé and Quai du Commerce, with trams now running at the extreme edges of the highway. This was the last city centre service to use 7000-type cars. Road and metro tunnels lie underneath, and there is an interchange here at Ribaucourt station from what is now service 51, Heysel to Van Haelen.

Had line 2 survived as a pre-metro operation it was expected that a number of through services would have continued from the north-west of the city where there were several upgraded lines such as Bvd de Smet de Mayer in Jette, now used by service 51, as well as inter-suburban service 19 through Simonis.

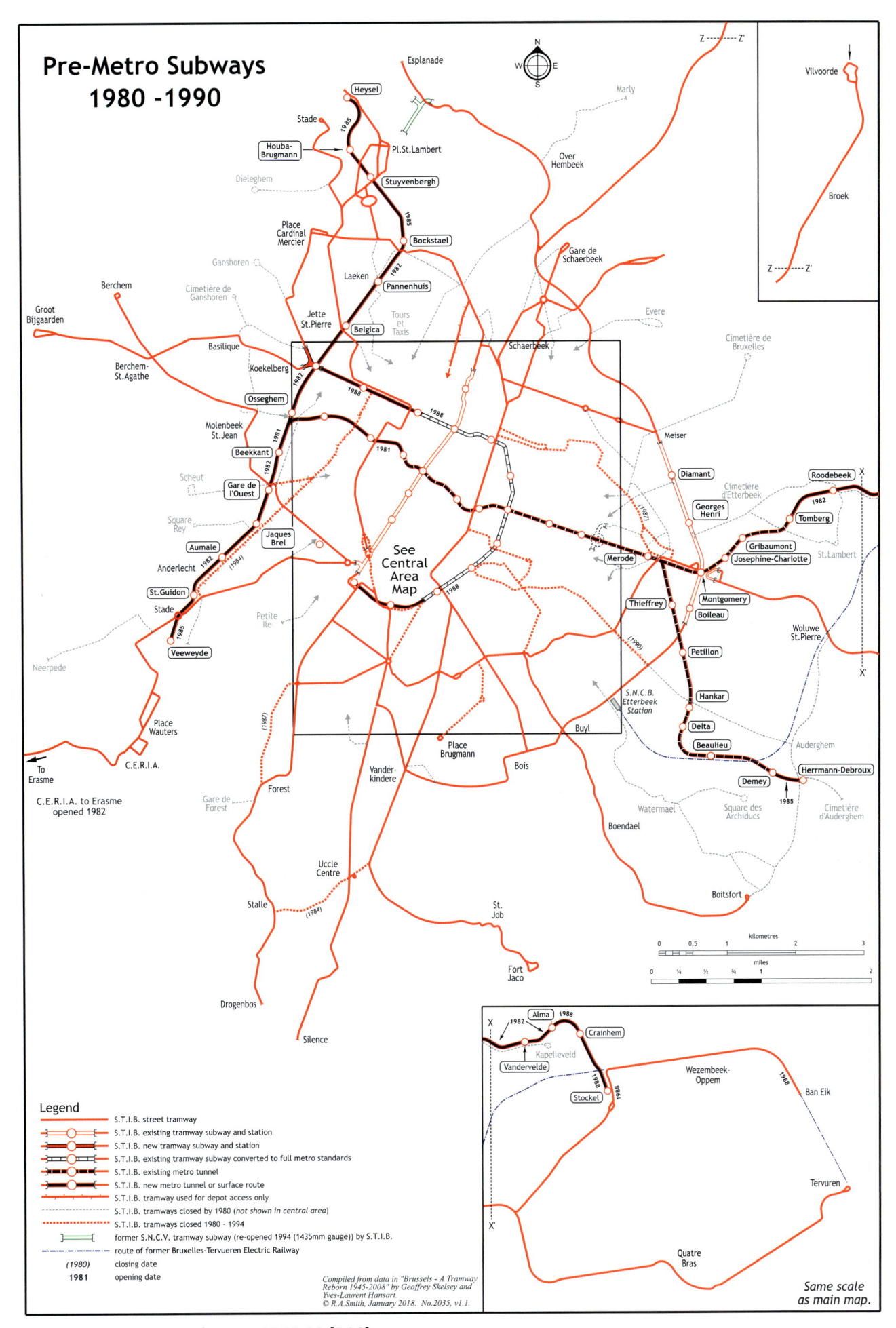

Map 21. Pre-metro and metro 1980-90 [RAS]

to present conversion of the line to Metro as of wider public benefit in three respects. It would provide faster trains and additional capacity which would benefit all passengers. It would need no additional expenditure on rolling stock as there were sufficient spare cars in the existing fleet. And by eliminating interim expenditure on short-term expedients such as pre-metro signalling, current collection, and platform configuration, tax-payers' money would be saved. A better service would be provided at lower cost. The arguments were debatable, and were bitterly opposed by many in Brussels, but De Croo was not sympathetic to what he would have regarded as sectional interests and his plan went ahead.

The arguments for and against it were many and varied. The principal grievance was the discontinuation of through one-seat travel to and from the seven remaining tram lines which fed into the tunnel, passengers being required to change at each point of intersection.

Despite a vigorous campaign the plans proceeded as announced and tram 7912 was the last to run on this portion of the Ceinture on the evening of 27th September 1988. Over four succeeding days the huge task was undertaken of altering platforms and signalling, adjusting and commissioning third-rail power supply and metro signalling, and removing the temporary ramps at Porte de Hal, Rogier, and Simonis. Five new stations were inaugurated, at Porte de Hal, Hotel des Monnaies, and Gare du Midi (deep level), and on the northern sector at Yser, and Ribaucourt (a station planned at Square Sainctelette close to the Charleroi Canal was not built). Despite valiant work by the engineers the early days of the new service were marred by some breakdowns and interruptions.

At Simonis the existing pre-metro station was converted for metro use and a new sharply-curving tramway station was built above to the west, enabling a new feeder tram service 19 to connect with the metro: this has become one of the busiest lines on the tram system. At street level tram tracks were retained between Quai du Commerce and Boulevard du Jubilé, enabling a remaining tram service (now No 51) to operate on relocated tracks along this short stretch of the Ceinture, connecting with the Metro at Ribaucourt and running down the western side of the Pentagon to the Midi tunnels at Lemonnier.

From 1988 the Metro trains ran into the deep-level station at Gare du Midi, but for the time being trams serving St-Gilles and beyond continued above ground along a stretch of the Ceinture as far as Chaussée de Waterloo. As we have seen these continued until the line 3 extension to Albert was opened in 1993. The further extension of the line 2 Metro beyond Midi will be described later, but exceptional constructional problems delayed its completion until 2009, followed by accompanying recasting of the whole metro service.

6.3 Type 10,000: a lost opportunity

Hybrid rolling stock, sometimes termed 'Stadtbahn cars', originated in Germany in 1964-66, as a means of addressing exactly the issues identified in Belgium a decade or so later, that of wedding a modern underground system with high platforms to through operation over suburban lines on which not all stops could be so equipped. Over a hundred such cars were supplied to Frankfurt-am-Main. They were six-axle articulated vehicles, most with folding steps, about 24m long and 2.65m wide, taking about 170 passengers per set. Similar arrangements with folding steps were adopted in Hannover and elsewhere.

It was realised in Brussels by 1978 that systematic development of comprehensive light-rail services linked to the pre-metro network would require new types of rolling stock. In keeping with the Frankfurt developments this would take the form of hybrid low platform/high platform stock capable of operation in underground stations with easy and rapid access, but also adaptable to street-level boarding on conventional tramway sections. To achieve this folding steps were required. The estimated requirement for system-wide operation was 235 cars.

These would have been eight-axle vehicles, 31.5m long (somewhat longer than the 7900 series), and of standard tramway width (2.20m) for compatibility with existing infrastructure. Installed power (8 x 73 kW motors) would be substantially greater than for existing stock and would therefore offer improved acceleration and speed on segregated sections. Multiple-unit couplers were specified, and the cars would have been of distinctive, not to say eccentric, external appearance. In 1981 fourteen initial cars were ordered from BN, for use on line 2 and for the elimination of the small 7000-types. The unit cost would have been almost twice that of the 7900-type, and the order was cancelled a year later in the atmosphere of austerity espoused by the central government.

In the longer term this problem was met more satisfactorily by the use of low-floor stock, but this would not be practicable for many years.

6.4 Line 3: studies in 2001

When Minister de Croo formally opened the new Louise tram subway station on 19th August 1985 (there is still a rather battered commemorative plaque in the concourse) it seemed that progressive extension of the Ceinture pre-metro was proceeding on course, which made his sudden announcement little more than a year later, that the whole route would be completed as quickly as possible as a Metro, all the more of a shock. There can be little doubt that, had his policies been allowed to continue the remaining central area pre-metro, line 3 between Nord and Midi, would have been next on the list. Long-term metro plans evolved over

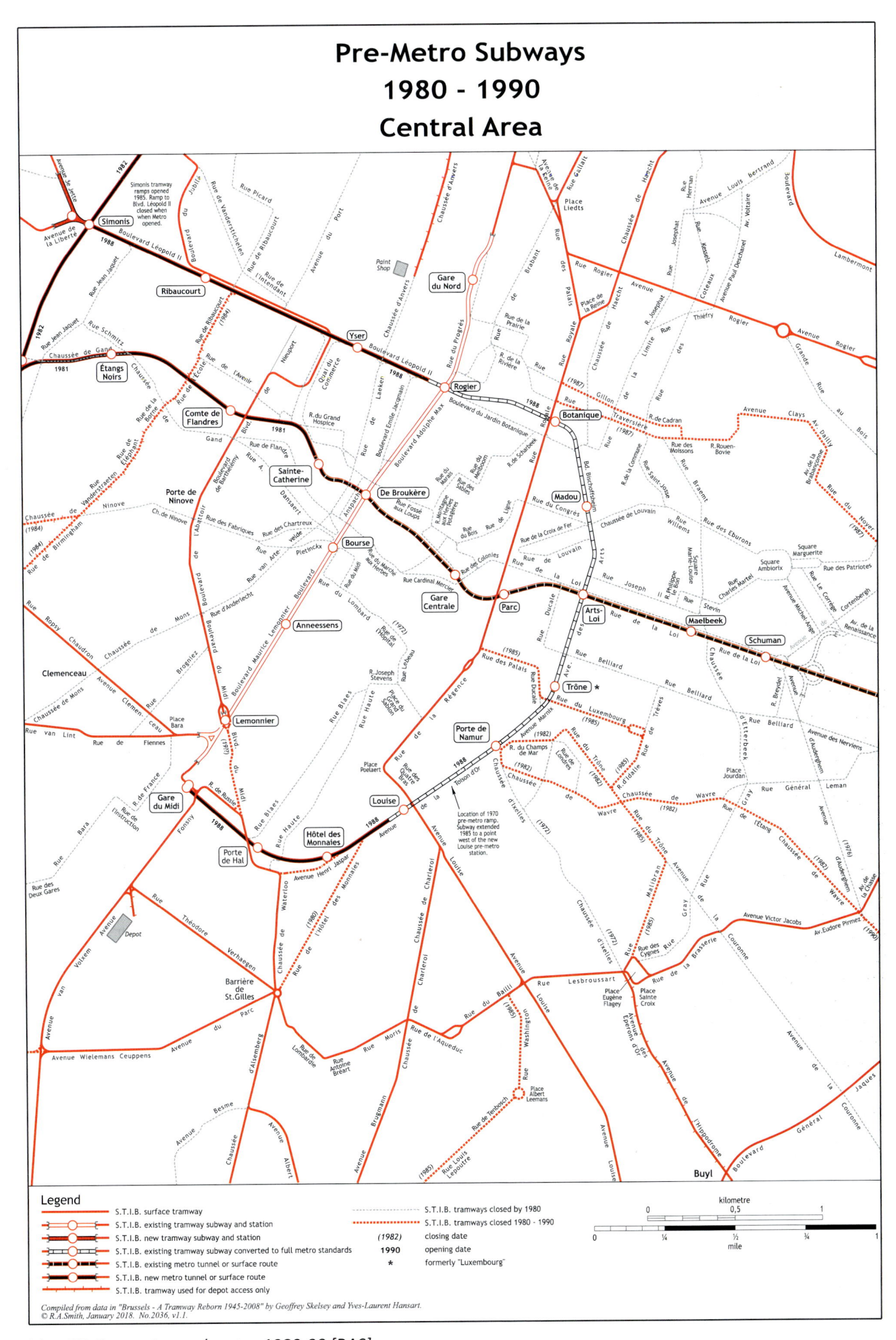

Map 22. Pre-metro and metro 1980-90 [RAS]

Amongst measures taken to improve the quality of service on line 3 was the commissioning for tramway use of the unused centre platforms at stations between De Brouckère and Anneessens, allowing lateral passenger flow and lessening dwell times. This is De Brouckère in December 2002.

North of Gare du Nord narrow streets and awkward layouts threatened service standards, especially of service 55. One of the pinch points was at the remarkable circular layout at Place Verboekhoven ('Cage aux Ours' or 'bear pit), where trams skirted a railway cutting. Seen in 1984 the conflict is all too clear, and although this layout has been much improved the general problem remains. A new station on the extended line 3 Metro is planned near here.

South of Albert, where the 1993 tunnels ended, some streets were narrow and often congested, like Rue de l'Etoile south of Uccle. One object of the changes announced in 2003 was to concentrate on line 3 those services least exposed to traffic, so long-standing cross-city service 52 was curtailed and only its northern section ran through the tunnel.

time, and there were many changes of mind, but a line between Uccle Globe and Schaerbeek Station, incorporating the north-south tram tunnel, seemed a likely proposal, with the possibility of a surface extension north-westwards to Esplanade (although even then this would have been likely to have raised formidable environmental objections). That this didn't occur for many years is a consequence of the radical restructuring of Belgian governance, and the agreement in 1990 of a new relationship between the STIB and the autonomous Capital Region. Transport planning and the STIB's obligations and entitlements, changed profoundly and began to reflect local aspirations. It is no coincidence that these events ushered in a revived role for the tramways, under the notable leadership of M J-L Thys, Minister of Communications in the new Regional Government, and former Mayor of Jette. He was a tireless advocate of the tramways until his premature death in 1999.

Having said this, tram operation of the north-south pre-metro was generally recognized to be far from satisfactory, and as traffic grew substantially in the 1990s under the influence both of large-scale office development around Midi and Nord and the consequences of the southwards extension of the line into new areas in 1993, this became all too apparent. There were two fundamental problems which prevented the precious asset of the tunnel and its stations fulfilling their potential. As elsewhere, the then-existing tram rolling stock was unsuitable, with limited capacity and awkward access leading to extended dwell times. And the service pattern, the result of years of haphazard evolution, was far from optimal.

The five services using the subway were largely the descendants of those which had used the surface lines before 1976. The predecessors of service 52 (then Esplanade and Drogenbos) had operated since 1935; the 81 (by then, Heysel – Gare du Midi – Montgomery) had commenced in 1914. They did not necessarily meet present-day requirements. These routes were all, to a greater or lesser extent, exposed to street traffic congestion which led to delays and bunching. Finally, they were timetabled to different frequencies, meaning that headways in the tunnel were irregular.

Some corrective measures were relatively easy to implement. Between 2002 and 2004 the unused centre platforms at the four intermediate stations were adapted for tramway use, with lowered sections and new access arrangements. This allowed lateral loading and unloading with beneficial results for dwell times. At Gare du Nord, which had been designed as a four-platform interchange, the two unused lines were brought into use in November 2004 to improve traffic flow. A new turn-back siding was installed just outside the northern portal to replace reversal within the station. These were palliatives and in 2001 the Capital Region commissioned a thorough appraisal of the tunnel and its services with a view to a long-term plan. There were several possibilities, including:

The service selected for through operation along the line 3 tunnel was former route 52, extending south from the Esplanade terminus (built for the 1958 Exhibition and seen here with one of the first 3000-type cars in 2008). This used almost exclusively segregated tracks, and became new service 4.

• A limited Metro service between Nord and Albert, which would have required complete re-building of the Constitution tunnels, but for which most of the infrastructure otherwise existed. Feeder tram services would continue from the north into the terminal bays at Rogier, and from the south into a new interchange at Albert for which provision had been made when the station was built. Connections with line 2 at Midi would provide depot and service access to the line. This would have improved speed and reliability on the core route but would have required most suburban passengers to change modes at least once.

• A longer Metro operation, extending northwards from Nord to Haren (where a depot site was available) and south, possibly as far as Uccle.

• A more radical plan, using tram-train technology to link with SNCB services at Midi and Nord, providing better direct coverage of the Region and relieving the overloaded Junction railway.

• Continued tramway operation but using new and more suitable rolling stock and a radical restructuring of routing and operation to improve reliability and capacity.

In mid-March 2003 the results of the inquiry were announced. Tram operation would continue for the foreseeable future, but the line would become a 'quasi-metro' with a high frequency daytime shuttle

Fine reserved track tramways led eastwards out of Albert station, and the new services 3 and 4 made best use of these. This is Avenue Winston Churchill.

A similar wintry scene along Avenue Albert east of the tunnel portal. Uninterrupted tram traffic was possible on these and similar lines.

service covering the core route between Nord and RP Churchill (service 3). The 8-axle low floor cars (type 4000) then in course of delivery would be dedicated to this service and would offer substantial improvements in accessibility and capacity. A single day-time through tram service would continue on the whole subway section (originally service 4), replacing venerable service 52 and running between Esplanade and Parking Stalle. Together with improvements to the surface tramways, notably the new line along Quai des Usines opened in 2006 which avoided a congested area, this route would offer higher standards of speed and reliability. New service 4 ran almost entirely in tunnel or on segregated track, a long-delayed fulfilment of one of the original pre-metro principles.

These improvements would be achieved at a fraction of the cost of conversion. These changes were broadly implemented in 2006-7, partly with a different service pattern in the evenings offering a wider range of suburban connections to and from the centre.

From 31st August 2009 the service was adjusted and lines 3 and 4 changed their northern termini. Service 3 now runs between RP Churchill and Esplanade, while service 4 was shortened to overlap with it and to run between Parking Stalle and North Station. Today services 3 and 4 offer a combined 24 cars an hour operation in each direction during the day over the core section of line 3. However the section south of Gare du Midi was still affected by surface delays: service 51, now running between Heysel and Van Haelen via the west side of the Petite Ceinture and into the tunnel at Midi, had several pinch-points on its route. Similarly passengers from the north-east using service 55 (now running between Da Vinci and Rogier) were subject to delays and some had to change to complete their journeys.

It became clear, despite improved service on the main north-south axis, that other problems remained. The awkward arrangements within the 1957 tunnels at Midi, with their conflicting junctions and slow speed, constrained service levels. The need for a substantial proportion of passengers to change cars to access the tunnel stations was unpopular. Most seriously the 23 service, the busiest on the system, which had joined the east side peripheral line with the pre-metro, was severed and most of its passengers were required to change to the shuttle at RP Churchill. This would not have been so bad had the arrangements there not been makeshift. The STIB argued that faster and more reliable service compensated for the inconvenience of a change but public dismay never fully subsided. Some services were still subject to traffic delays.

Continued traffic growth inevitably cast the future of the operation into doubt, and as we shall see its eventual conversion to Metro operation became increasingly likely.

A reminder that the pre-metro concept originally included high-quality suburban surface lines, like the Erasme Hospital extension opened in 1982. It was closed to trams in 1999, and became a metro route in 2003. Note heavy rail and ballast and catenary overhead line. [MJR]

7. Pre-metro in other Belgian Cities

In the expansive and optimistic 1960s both the money and the will were present, in Belgium and elsewhere, to carry out major infrastructure projects to overcome years of neglect during the Depression years, and in the War and its aftermath. As we have seen the (then) unitary Ministry of Communications agreed in 1962 to instruct local authorities and transport operators in Belgium's five largest urban areas to produce plans for new fixed-track transport systems, to supplement or replace existing tramways. In this chapter we look at the outcome of these intentions. Only in the cases of Antwerpen and Brussels was the pure concept of pre-metro technology adopted, and their subsequent policies later diverged. Two of the other cases (Gent and Liège) did not proceed any further, and in Charleroi a hybrid system emerged, incompletely and slowly. Before regionalization in 1988-9 the political dynamics of the invariable coalition governments required a certain measure of equalisation in allocation of capital resources between the two main cultural and language communities, and this is reflected in the outcome.

The Antwerpen pre-metro reaches out from the city, where possible on segregated track. This is the Melsele terminus on the left bank of the Scheldte, with a bus interchange, park-and-ride site, and bicycle hub.

7.1 Antwerpen
7.11 The City

Antwerpen is one of the great European cities, a centre of trade and culture for over a millennium, and is in our time the second busiest port in Europe and a focus for varied and successful industries as well as the ancient and lucrative craft of diamond polishing and trading.

The city stands mainly on the east bank of the River Scheldt, and had horse tramways from 1873 with several different operating companies and no standardisation, even of track gauge. These lines lay mainly within the lines of the fortifications, which had been renewed in the mid-nineteenth century. In 1899 most of the undertakings merged into the *Compagnie Générale des Tramways d'Anvers* [sic], which introduced new standards including adoption of metre-gauge track (suited to narrow streets in the old city). Electrification began in 1902 and was completed the following year. In 1927 a new franchise let the undertaking to *Les Tramways d'Anvers*, succeeded in 1946 by an interim public-private organisation *Tramwegen van Antwerpen en Omstreken* (TAO). Like the other cities considered here Antwerp also had an extensive suburban and interurban Vicinal network, all on metre gauge after 1921. Sadly economics, and the need for investment in track renewal and relocation, led to the network closing between 1959 and 1968.

Belgium's difficult political and economic circumstances between the Wars had prevented any investment in new urban rolling stock and by the 1950s the Antwerp fleet consisted of a mixed inheritance of 192 relatively small two-axle cars, some over fifty years old, mostly operating with trailers. Some reconstruction was completed, including fitting of platform doors and improved seating, but the long-term future of the network was doubtful, and several lines were converted to diesel bus operation between 1950 and 1964; there was also a limited flirtation with the trolleybus between 1929 and 1964. Nonetheless TAO, encouraged by the favourable results of PCC-operation in Brussels from 1952, especially in reducing costs by eliminating trailer operation, ordered a first group of 39 PCC cars in 1959. They began to enter service from 1st December 1960 by when 22 further cars had been added to the order, with many more to follow in three further batches. Antwerp eventually had the largest West European fleet of virtually identical PCC cars.

In common with other Belgian city tramways a further administrative re-organisation took effect in 1963, combining public and private elements and creating the *Maatschappij voor het Intercommunaal Vervoer te Antwerpen* (MIVA), which lasted until the Flemish regional operator De Lijn took over in 1991, after regionalisation.

The narrow streets in the old centre of the city produced intolerable congestion, and made the Ministry's plans for pre-metro tunnels attractive. However, outside the centre, broad avenues, including some occupying the sites of former fortifications,

Until the late 1950s the Antwerpen system was operated wholly by small traditional four-wheelers with trailers. This evocative scene from 1952 captures the essence of Belgian life in that distant era. [Courtesy Patrick Sellar]

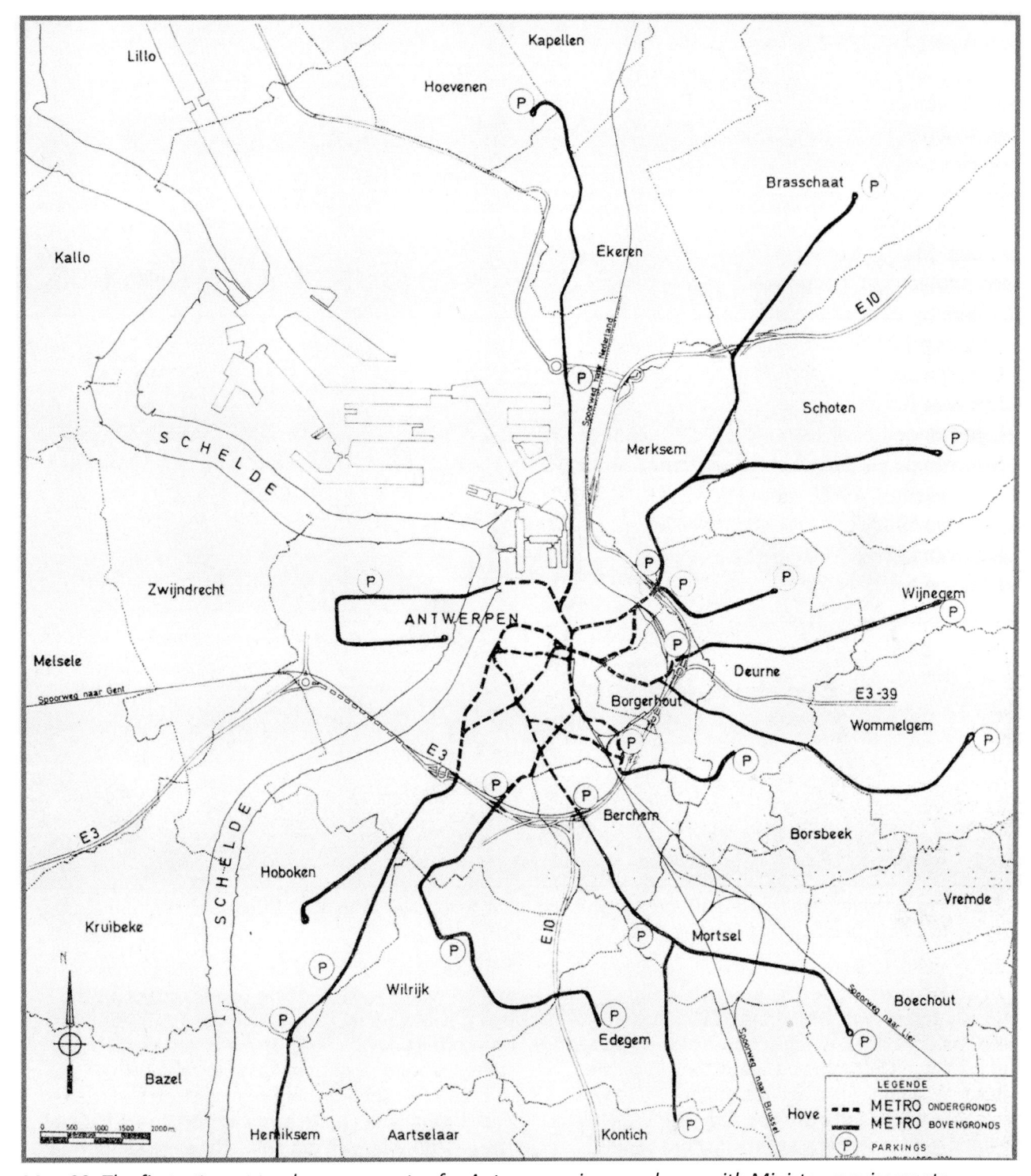

Map 23. The first attempt to plan a pre-metro for Antwerpen, in accordance with Ministry requirements, produced this complex inner-area tunnel system replicating existing surface lines; it would have been costly to build and difficult to operate.

made it possible to build substantial stretches of tramway reservation, and compared with Brussels today the tunnelled sections are largely confined to the central area. The underground network remains a classic pre-metro operation, with multiple suburban services entering and leaving the tunnels as appropriate to provide uninterrupted travel to or across the centre. We must recall that, as set out in the 1960s, the concept was not confined to tunnelling: upgrading of suburban surface lines was a crucial part of the project, and was partly achieved in Brussels (such as services 3 and 7 today). Developments in Antwerpen have carried the model still further.

7.12 National Plans and initial response

It will be recalled that the central Ministry of Communications set up a nationwide planning body in January 1962, *Dienst Bevordering van het Stedelijk Vervoer* (BSV/PTU), to coordinate a programme of improvements to urban public transit. A local body in the five main cities (in *Antwerpen Bijzondere Studiedienst*, BS) was charged with drawing up appropriate plans.

Map 23 shows the initial response of the Antwerpen BS to the Ministerial invitation. Published in 1962

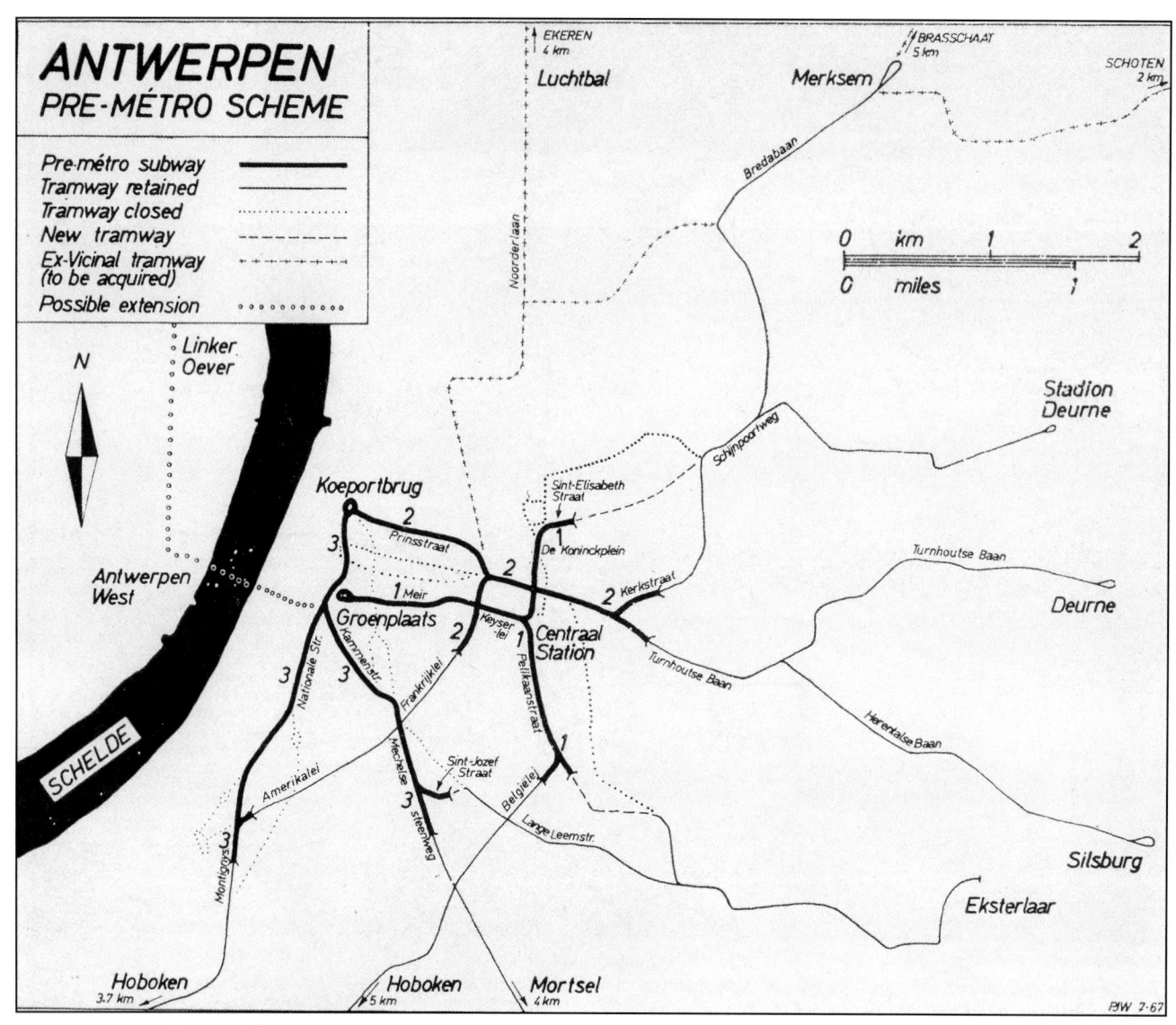

Map 24. A much-simplified and practicable scheme had been devised by 1966, with three separate tunnel axes. Lines 1 and 2 have been partly built, line 3 has not.

it represents in large measure an underground alternative to existing surface tramways, with a dense central network including many junctions and connecting lines. In effect about eleven radial lines converged into tunnels below the old centre, with some running links between them. There were many objections to this project. Although the vertical separation of tracks meant that junctions could be non-conflicting divergences were so frequent and closely-spaced that they were certain to have a damaging effect on reliability. The provision of so many separate lines in a small area meant that the concentration of traffic needed to justify the huge costs of tunnel and station construction would not arise. And construction below the majority of principal streets would be long-drawn-out and intolerably disruptive. In the event almost none of these proposals was proceeded with and a second response was issued five years later.

The revised 1967 proposals, partly implemented from 1975, can best be understood in terms of three separate overlying axes. Map 24 will help to follow the descriptions. Axis 1 was to reach eastwards from an underground turning loop below Groenplaats,

running below Meir and De Keyserlei as far as Central Station where it divided into two branches (as the two tracks were laid one above the other this was a non-conflicting junction). There would be separate stops serving the station, Astrid to the north, Diamant to the south. The northern branch stretched to De Konickplein with an exit ramp in Sint-Elisabethstraat. The southern branch ran beneath Lange Kievitstraat, then diverging to two exit ramps into Belgielei and Mercatorstraat. This axis amounted to 3.7 route/km.

The second axis would have been further north, running east from another underground loop at Koepoortbrug below Ossenmarkt and Rooseveltplaats, where it was joined in a trailing junction with a line running from the south via Frankrijklei and Opera. It then continued eastwards beneath Carnotstraat, diverging into two branches and exit ramps at Kerkstraat and Turnhoutsebaan. This line totalled 3.4 route/km.

The third proposed axis ran north-south, beginning at a second turning loop at Koepoortbrug below that for the second axis, with connecting spurs. South from here the line passed below Groenplaats (without any connection to line 1) and then immediately diverge into two tunnels below Nationalestraat to a ramp at Amerikalei, and below Kammenstraat to a ramp into Montignystraat.

At this stage, it should be noted, the intentions closely mirrored those operating in Brussels after 1969. The first stations were built with 95m platforms capable of accommodating metro trains, and the tunnels were built to standard-gauge dimensions, a condition for grant aid from government. However, the Brussels expedient of lowered sections in the platforms was not adopted and instead track was 'temporarily' raised in the station areas to permit tramcar boarding. No date was set for eventual conversion to metro operation.

7.13 Pre-metro as built, 1975 - 2017

Execution and adaptation of these plans was long and uncertain, and what was eventually achieved differed from them in important respects. Excavation began in February 1970 at Opera and proceeded largely by cut-and-cover, made considerably more intrusive by the superimposed track arrangement. The initial 1.23km between Groenplaats and Opera was eventually completed and opened in 1975 (the table shows the chronology of completion). Opera station was built on three levels to accommodate future lines 2 and 3 (above). Line 1 was extended further south-east in 1980 as far as Diamant and Plantin stations.

Work at Astrid station near Central Station began in 1975 with tunnelling proceeding towards the north-east. The near intolerable disruption caused by cut-and-cover excavation in the principal commercial streets led to a change in technology and much of the remainder of the system has been cut using boring machines. This north-easterly branch was extended considerably further than originally intended, and eventually reached 3.14km as far as a ramp at 'Sport' with four intermediate stations. Although main constructional work was completed the line was not fully equipped and remained unused.

At Central station a triangular layout was built, making through north-south running possible, but this had no public service until 2007.

On line 2 a substantially altered and more extensive itinerary was begun, with a line stretching (as planned) north from a loop at Opera (deep level) and then eastwards towards Carnotstraat, where a north-easterly line diverged towards Schijnpoort on line 1; and an easterly prolongation towards Zegel. Although largely complete these sections remained unused for several years.

The next line to be built was not foreseen in the 1967 plans although it had been a previous intention. Part of the city of Antwerpen lies on the western (left) bank of the River Schelde, then linked to the centre by a road tunnel and an intensive ferry service. In 1983 work began on tunnelling below the river, for a 1.5km extension of line 1 beyond Groenplaats to the west bank, where it emerged at a ramp and ran as a

The first PCC cars, seen here in original livery but with one-person operation, brought new standards to the undertaking. We are looking westwards in 1970, away from Central Station and towards the Cathedral in the far distance. These streets would shortly be excavated for the first section of the pre-metro tunnels. [Courtesy John Bromley]

In the Groenplaats not far from the Cathedral a PCC passes the statue of one of Antwerpen's most celebrated sons, the painter P.P. Rubens. It is running on temporary track on the edge of excavations for the tram tunnels. [MJR]

Another Groenplaats view indicates the disruptive scale of the tunnel work, but also the extent of the underground station which was built on three levels to accommodate future line 3 (which has yet to be built). [MJR]

Before tunnels were built beneath the river an intensive ferry service operated between Linkeroever and Antwerpen centre, with an electrified Vicinal line running in from Hamme in connexion. This closed in September 1959 and the replacement buses ran through the tunnel to the city centre. The Cathedral spire can be seen on the far bank in the left background.

Massive excavation returned to central Antwerpen in 2017, with consequences for business and traffic circulation: this is Kipdorpbrug on what will be the northern extension from Opera towards the former docklands. [MJR]

In 1990 the Antwerpen pre-metro was extended from the old Groenplaats terminus beneath the river to the left bank. This PCC in an older livery is climbing the ramp at Linkeroever. Just above the signal lamp can again be seen the spire of the Cathedral. [MJR]

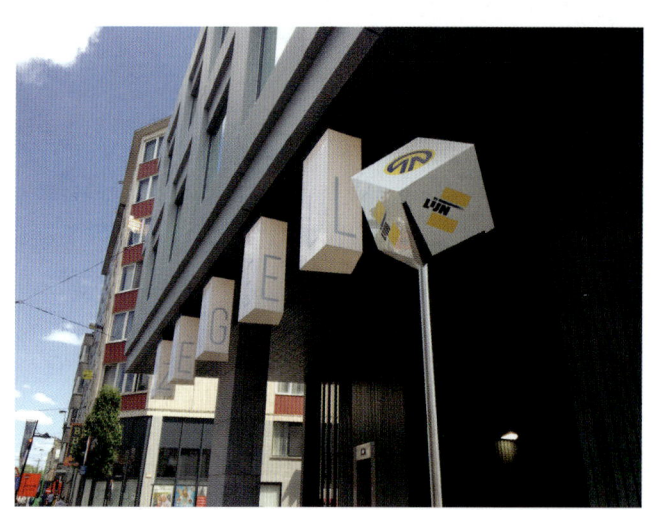

De Lijn devised an attractive 'tunnel' symbol, seen at Zegel station in 2017.

The opening in June 2017 of the southern tunnel from Opera towards Stadtspark included this classic subway ramp leading into Frankrijklei, used by services 8 and 10. [MJR]

A street festival accompanied the opening of the Wommelgem extension, with a fine display of historic trams including No 200, built by Franco-Belge in 1901, representing the first era of electric trams in the city.

On 15 April 2015 the long-mothballed 4km line below Turnhoutsebaan was commissioned for service 8 to Wommelgem P & R. The nickname 'Reuzenpijp' ('Giant Pipe') was coined, because the section consists of superimposed tubes. Much was made of the fact that transit between the parking site and the city centre took only 15 minutes.

modern surface tramway, originally to Linkeroever in 1990, and in 2002 to Zwijndrecht. As the sole high-capacity public transport cross-river link this has proved to be a busy and lucrative line. The under-river tunnel was built to standard-gauge dimensions.

Meanwhile the rest of the pre-metro project gathered dust. Financial constraints, especially after regionalisation which brought central government capital grants to an end, imposed severe new restrictions, and work on all but the Scheldte tunnels was suspended in 1988. Several kilometres of completed tunnel structure were mothballed, included the long north-easterly branch from Central Station to Sport which at last opened after nearly ten years of disuse. The line towards Carnot remained idle still longer, and its easterly section is still unused as are several intermediate stations.

In October 2007 a service was added using the third side of the triangle at Central Station, between Diamant and Astrid, and in September 2012, a further new service was introduced, using the previously idle southern ramp south of Plantin station.

The line from Astrid towards Foorplein finally opened in April 2015, but the stations at Carnot and Drink have not been fitted out and only Zegel is served. This route was also extended in June 2017, at its inner end, further south from Astrid surfacing at new ramp into Frankrijklei, completing one of the branches of line 2. The interchange station at Opera will not be completed until 2018, however.

In brief therefore, to summarise the situation in 2017: line 1 has been completed partly as planned, with a westerly extension below the Schelde and inner-suburban branches north and south from Central Station. Line 2 is partly complete between Opera and Schijnpoort. Its branch from Carnot towards Foorplein

Antwerpen Pre-Metro style.

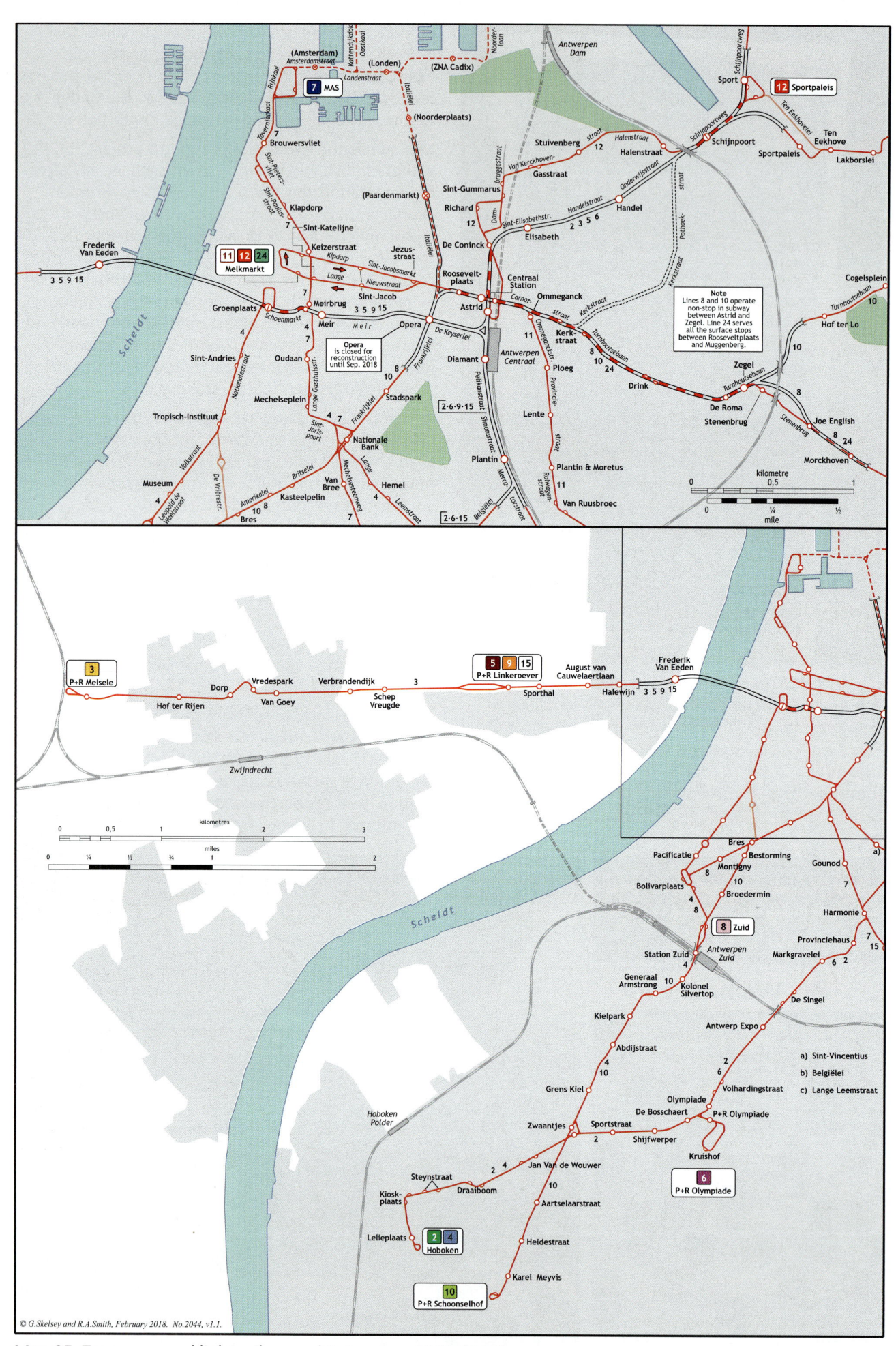

Map 25. Tramways and light rail around Antwerpen 2017. [RAS]

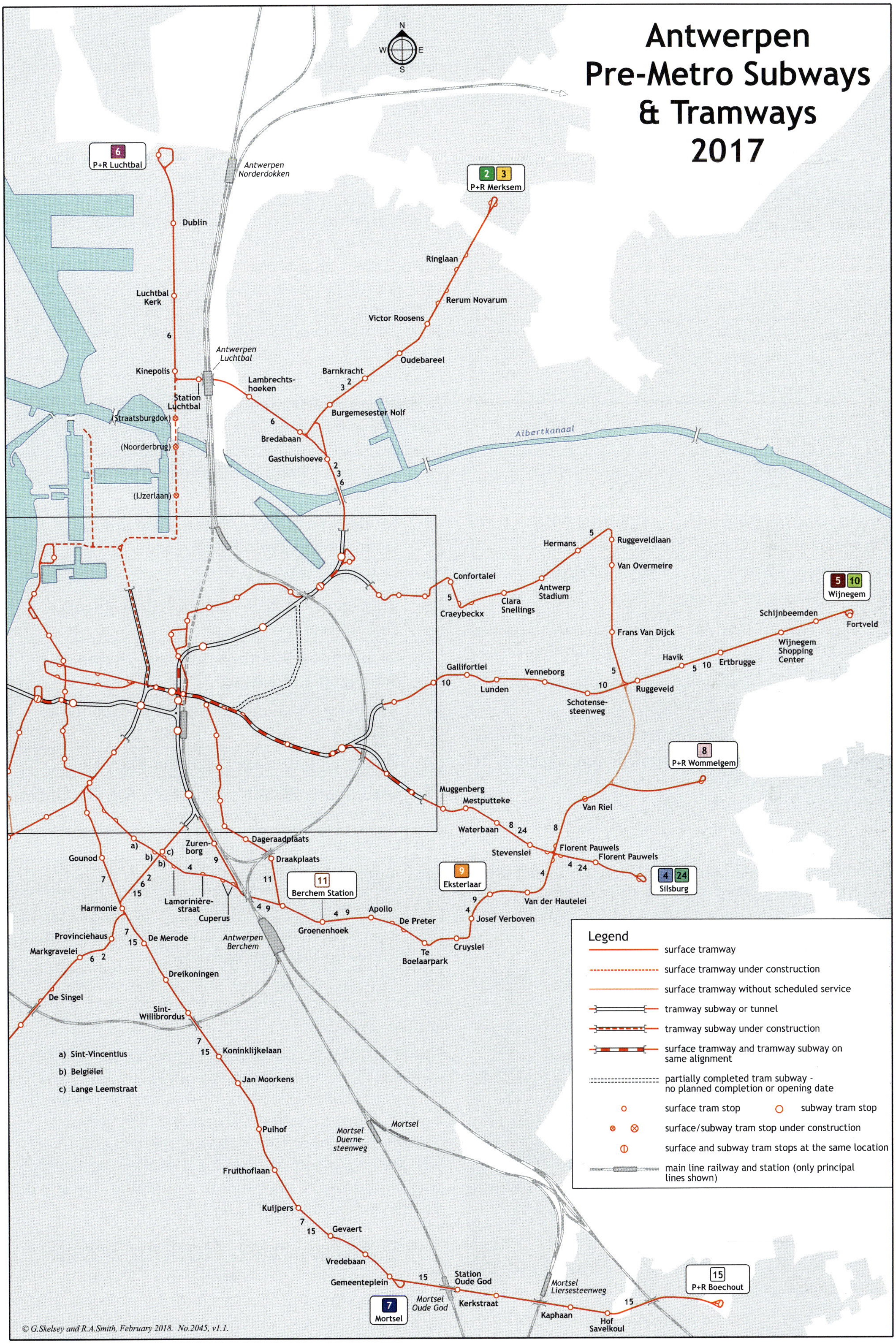

Antwerpen
Pre-Metro Subways & Tramways
2017
N
W
E
S
6 P+R Luchtbal
Antwerpen Norderdokken
2 3 P+R Merksem
Dublin
Ringlaan
Luchtbal Kerk
Rerum Novarum
Victor Roosens
Antwerpen Luchtbal
Oudebareel
Kinepolis
Barnkracht
Lambrechts-hoeken
Station Luchtbal
(Straatsburgdok)
Burgemeester Nolf
Bredabaan
(Noorderbrug)
Gasthuishoeve
Albertkanaal
(IJzerlaan)
Hermans
Ruggeveldlaan
Van Overmeire
Confortalei
Antwerp Stadium
5 10 Wijnegem
Clara Snellings
Craeybeckx
Schijnbeemden
Fortveld
Frans Van Dijck
Wijnegem Shopping Center
Havik
Ertbrugge
Gallifortlei
Venneborg
Lunden
Ruggeveld
Schotense-steenweg
8 P+R Wommelgem
Muggenberg
Mestputteke
Van Riel
Waterbaan
Florent Pauwels
Florent Pauwels
Stevenslei
4 24 Silsburg
Zuren-borg
Dageraadplaats
Gounod
Draakplaats
9 Eksterlaar
11 Berchem Station
Van der Hautelei
Lamorinière-straat
Harmonie
Cuperus
Apollo
Josef Verboven
De Preter
Groenenhoek
Provinciehaus
De Merode
Antwerpen Berchem
Markgravelei
Cruyslei
Te Boelaarpark
Dreikoningen
De Singel
Sint-Willibrordus
a) Sint-Vincentius
b) Belgiëlei
c) Lange Leemstraat
Koninklijkelaan
Jan Moorkens
Mortsel Duerne-steenweg
Mortsel
Pulhof
Fruithoflaan
Kuijpers
Gevaert
Vredebaan
Station Oude God
Gemeenteplein
Mortsel Liersesteenweg
15 P+R Boechout
7 Mortsel
Mortsel Oude God
Kerkstraat
Kaphaan
Hof Savelkoul
Legend
surface tramway
surface tramway under construction
surface tramway without scheduled service
tramway subway or tunnel
tramway subway under construction
surface tramway and tramway subway on same alignment
partially completed tram subway - no planned completion or opening date
surface tram stop
subway tram stop
surface/subway tram stop under construction
surface and subway tram stops at the same location
main line railway and station (only principal lines shown)
© G.Skelsey and R.A.Smith, February 2018. No.2045, v1.1.

is now in use, but with only one intermediate station. A number of additional ramps have been brought into use to facilitate further access from the surface system to the tunnels. Other shorter completed sections remain unused. Numerous studies have been carried out, including 'Masterplan' in 2005 which produced the openings carried out in 2015-17.

Opening of Antwerpen pre-metro sections

Itinerary and length	Date of opening
Groenplaats – Opera (1.3km)	25 March 1975
Opera – Diamant – Plantin (1.5 km)	10 March 1980
Groenplaats – Frederik van Eeden (west bank) (1.5 km)	21 September 1990
Astrid – Sport (3.2 km)	2 April 1996
East side of triangle at Central Station	October 2007
Ramp at Plantin towards Zurenborgstr (service 9)	September 2012
Opera – Astrid (deep level) – Carnot – Schijnpoort (2.7 km)	18 April 2015
Opera – Franrijklei, ramp towards Stadspark (services 8, 10)	3 June 2017
Ramp at Foorplein towards Hof ter Lo (service 10)	16 September 2017

Source: *T2000 De Lijn/Antwerpen*

By mid-2017 there were about 11 km of pre-metro tunnels, with 12 stations. The latest opening, in September 2017, was another short stretch of tunnel from Zegel to Hof ter Lo (service 10) with a new ramp eastwards at Foorplein joining to existing reserved track. North of Opera station, which was closed for rebuilding for a period in 2017-18, a further new tunnel will stretch northwards to open up the Noorderlijn project serving the former dockland area (see map 25).

In a fine example of pre-metro planning the west bank line has been extended on segregated track to an intermodal hub at Melsele (Zwijndrecht).

7.14 Surface extensions

To a greater extent than in Brussels De Lijn has latterly pursued a consistent, rapid, and far-reaching policy of extensions to the surface system, linking to the pre-metro lines. Between 2012 and 2017 alone seven extensions, totalling more than 12km, were opened, including to two park-and-ride sites at Wommelgem and Luchtbal. An important future prospect involves opening up access to extensive new urban development in the north-west of the city on a former dock estate. By 2019 this 'Noorderlijn' will change the shape of the northern part of the network and extend its scope. The undertaking has published outline plans for a continuing programme of new lines through to 2025, including further extension on the left bank.

Antwerpen tramway services in June 2017
(Services in **bold type** use pre-metro lines)

2	Hoboken – Plantin – Diamant – Astrid - Merksem
3	**Merksem – Astrid – Opera – Groenplaats – Linkeroever - Zwijndrecht**
4	Hoboken – Silsburg
5	**Wijnegem – Astrid – Opera – Groenplaats – Linkeroever**
6	**Olympiade – Plantin – Diamant – Astrid – Schijnpoort - Luchtbal**
7	Mortsel – Sint-Pietersvliet (to be extended to Eijlande)
8	**Wommelgem – Zegel - Zuidstation**
9	**Eksterlaar – Plantin – Diamant - Astrid – Opera – Groenplaats - Linkeroever**
10	**Wijnegem – Schoonselhof**
11	Berchem – Melkmarkt
12	Sportpaleis – Melkmarkt
15	**Boechout – Plantin – Diamant - Astrid – Opera – Groenplaats - Linkeroever**
24	Silsburg – Melkmarkt

Source: *T2000*

Note that six different services use all or parts of the east-west line, fanning out at the eastern end to serve different corridors, a classic pre-metro pattern which existed in Brussels until 1976, and which provides a wider range of one-seat through services for passengers. As yet only service 8 uses the southerly line, and provides a fast link to a park-and-ride site on the line extension opened in 2015.

7.15 Antwerpen Rolling Stock

The undertaking had begun to replace its elderly fleet well before the Ministry initiative seeking pre-metro development, and they eventually acquired 166 almost identical single-ended four axle PCC cars, similar in most respects to the Brussels 1951 design.

Delivery of the first batch began in 1960, and they were a sensational sight in streets used only to aged two-axle trams.

Although the PCCs closely resembled the contemporary four-axle cars in Brussels there were differences arising from Antwerpen's operating practices. The centre doorway was situated in the centre of the nearside instead of being displaced one bay rearwards: this was to conform with the Antwerpen fare-collection system, with passengers boarding at the front and moving past a conductor seated just to the rear of the front doors. This also required a larger front platform, and the driver sat on the left-hand side, not centrally as in Brussels, and there was accordingly a blunter front end. The overall length of the cars was reduced by 18cm.

The cars had outside-framed metre-gauge B6 bogies, based on US originals used in, for example, Los Angeles. They were also equipped with pantographs from new. The second series of PCCs (originally numbered 2061-2100, 1965-6) were fitted from 1972 with Scharfenberg couplers for multiple-unit operation, and from the third series (2101-25, 1969) the conductors' units were omitted. The last series (2126-65, 1974) were equipped with couplers from new, and their delivery allowed withdrawal of the last two-axle cars. Antwerpen now had a uniform fleet. Twin-car operation began in 1973, a policy sadly omitted in Brussels.

One-person-operation was introduced from 1966, with a large blue diamond on the dash to indicate one-person cars. All cars were single-staffed from 1975. The electrical equipment of the PCCs was based on 1930s technology and beginning in 1975 trials were made of solid state controls, offering savings in current consumption and smoother operation. The third and fourth series were fully re-equipped in this way from 1989, and the bogies were also fitted with anti-sway dampers. The majority of the earlier cars were also overhauled from 1999.

The PCCs were unmatched in their riding qualities, and in their acceleration, but suffered from the same problems as the Brussels equivalents in subway operation, with slow loading up steep steps. De Lijn renumbered the fleet in 1991, in the series 7000 – 7165, pleasingly matching the sequence of the similar cars in Brussels.

Antwerpen PCCs

No of cars	Width	Length	Seats/Stand	Motors
166	2.20m	14.017m	30/75	4 x 55 kW

Despite the expansion of the system no new cars were added to the fleet between 1975 and 1999. There were attempts to interest the undertaking in replacement cars, notably a design drawn up by BN for an eight-axle version of their articulated cars built for the Coast and Charleroi from 1980. This came to nothing.

After full regionalisation in 1988-9 procurement became the responsibility of the regional transport operator *De Lijn*, which was set up in 1991 and managed tramways in Gent and on the Coast as well as in Antwerpen. When additional cars were at last acquired a standard regional design was adopted, although as Gent required double-ended vehicles there were necessarily variations. Antwerpen has retained universal single-ended operation, and although this offers advantages in terms of capacity, as well as initial and long-term costs, it also brings with it operational inflexibility in terminal working and emergency operation.

The first design of the modern era was a low-floor series built by Siemens and based on the Dresden NGT6DD model, one of which had been displayed in

With extension of pre-metro under Turnhoutsebaan in April 2015 a surface line reached a new P & R site at Wommelgem.

Running in a coupled pair this PCC set is about to depart west from Groenplaats and under the river to the terminus at Linkeroever. The blue diamond on the dash indicates one-person operation, but notice the cumbersome boarding arrangements.

Antwerpen operated its system with a uniform PCC fleet for 24 years, but manufacturers sought to promote vehicles more suited to subway operation. This is a sketch of BN's offering in 1980, of a hybrid car based on designs for Charleroi. Nothing came of this.

Interior of an Antwerpen PCC in original condition, with the conductor's desk in the left foreground. [Courtesy LRTA London Area/Online Transport Archive]

Antwerpen. Called in Belgium 'Hermelijn' the model ran on all three De Lijn systems, and the Antwerpen fleet of 84 five-module cars was delivered between 1999 and 2012 in four batches. They are 29.6m long and 2.30m wide, 10cm wider than the PCCs. Some of these cars have migrated to Oostende in successive summers, to handle booming holiday traffic there when demand is lower at home. Starting in 2017 Hermelijn cars were fitted with multiple-unit equipment, and entered service in that form in November. For their next order in August 2012 De Lijn returned to their faithful suppliers Bombardier and ordered a fleet of –initially- 28 of the 'Flexity 2' design, to be named 'Albatros'. They come in both five-section and seven-section versions. Given the extent of route extensions in recent years it remains to be seen whether the new cars will displace further PCCs. To meet short-term rolling stock needs some Gent PCCs have been transferred to Antwerpen late the end of 2017, where their double-ended capability would be useful.

7.16 Conclusion

The present Antwerpen operation is the best vindication in Belgium of the pre-metro principles set out over half a century ago, a comprehensively upgraded classic tramway which has provided tunnelled sections where needed in the centre and has energetically extended and improved its suburban network. Protracted vacillation, as in Charleroi, for a time discredited the project, as political issues and stringent economies moth-balled much of the part-completed works. In recent years a more determined attitude, as well as readier access to funding, has turned the situation round, and with new rolling stock the city now has an enviable and affordable light-rail network. Access to a network of park-and-ride sites is a particular feature of the undertaking. Although its origins are identical to those of Brussels lines 1 and 2 there can be little prospect now of a conversion to heavy metro ever taking place.

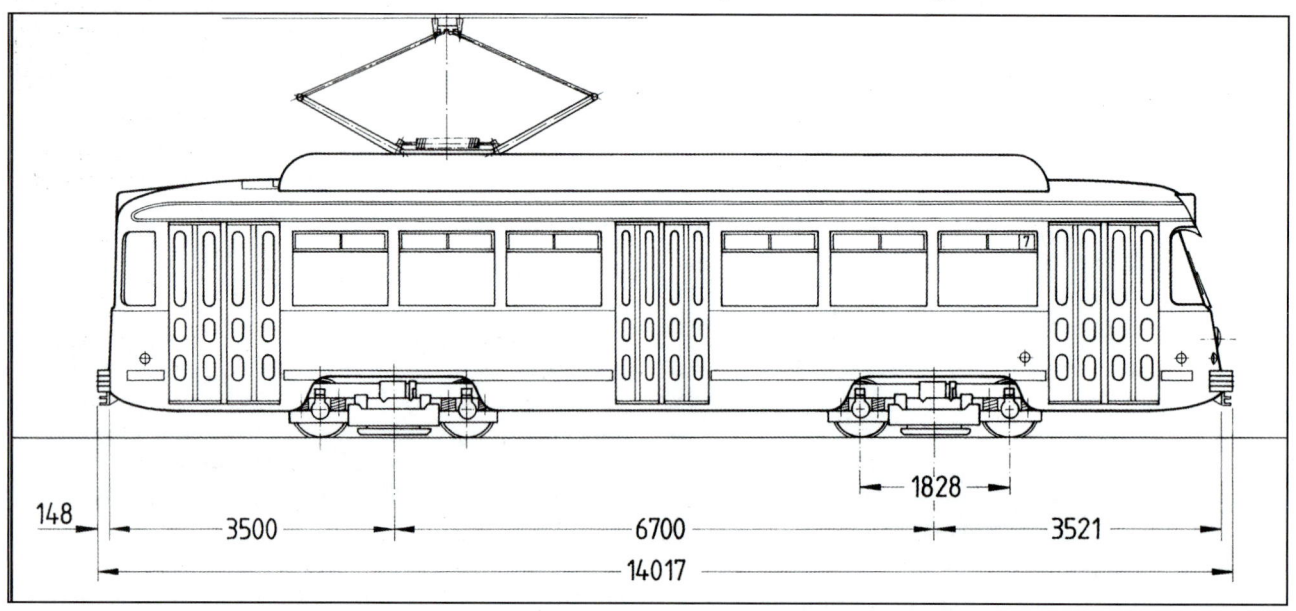

Diagram of the PCCs in original condition: the door arrangement and body shape differed from their Brussels contemporaries.

Gent's beauty is apparent from this view of a PCC in original livery crossing a city centre canal in 1984. It is easy to understand the resentment caused by plans to excavate tram subways here.

7.2 Gent

The Flemish Region city of Gent is a gem, one of the most beautiful sites in Europe with its gracious buildings, narrow streets, fortifications, and churches, all bordering placid canals. It is also a major commercial centre and inland port, with important academic and cultural institutions. When the national plans for Belgian public transport were launched the city's metre-gauge tramway system was a fascinating but outmoded survival of earlier times, with a fleet of pre-war cars and awkward track layouts. In most countries it would long before have succumbed to buses. Gent had also been a centre of a regional network of Vicinal lines, but these had all closed by 1959 (some of the routes have more recently been revived by extensions to the city system).

The city's size and growing economy encouraged the Ministry to include it amongst those meriting pre-metro development, and a series of proposals was produced of some inventiveness. They envisaged taking over several of the city's picturesque canals and using their courses as the core of a largely segregated light rail system, linked to a number of conventional subway sections and to existing and new surface lines. Map 26 shows the outline details.

Unsurprisingly these plans produced a storm of indignation at the prospective damage to the city's fabric and heritage. Construction would also have been exceptionally challenging. In consequence, the plans were abandoned and in 1971 a decision was taken to retain and refurbish most of the existing surface tramway and to order a fleet of double-ended PCC cars to operate it: eventually

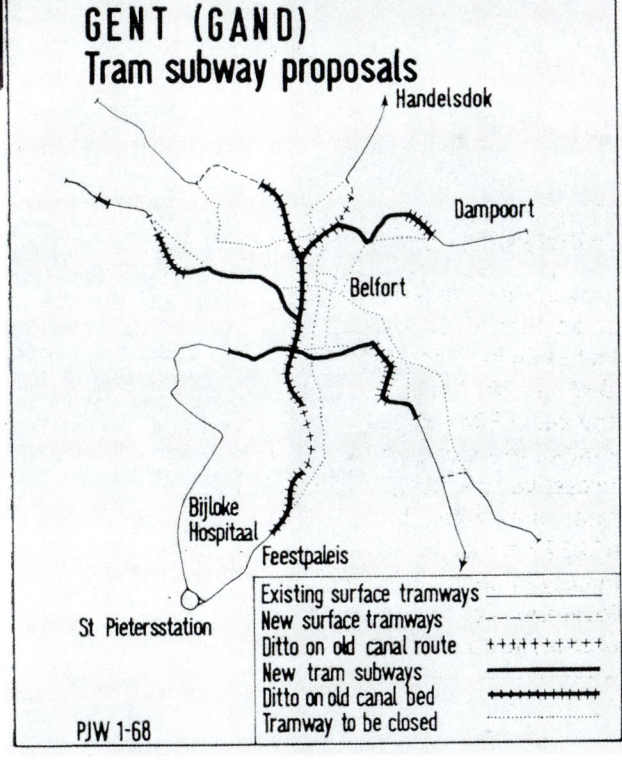

Map 26. Gent's 1968 proposals for pre-metro construction envisaged a net of tunnels replacing all inner-area surface lines, using canal beds for some of the construction. This was not popular.

54 cars were acquired. This was a brave and enterprising decision and was amply rewarded. Traffic substantially increased and eventually there were to be many additional extensions and a new low-floor fleet (the last PCC cars in normal service ran late in 2017, but a few were transferred for temporary use in Antwerpen). Significantly the retention of city tramways was accompanied by selective pedestrianisation, a world-wide trend which has improved regularity and accessibility without resorting to subways. Gent has therefore become, not an exponent of the pre-metro but of a cheaper, simpler, and more achievable type of tramway.

Before 1971 the Gent tramway was operated by a remarkable heritage of pre-war trams, many of three-axle design. Two pass in the main square, the Korenmarkt. The system's survival was miraculous. [Courtesy John Bromley]

By 1984 the Korenmarkt had been refurbished and the tram service had been taken over by a homogenous fleet of 54 double-ended PCCs. This is largely a pedestrian area now.

The Gent system was not only saved but has been transformed with extensions and a new tram fleet: one of the new Hermelijn cars is seen at the opening of the Maalte extension on 1 September 2004. No subways here though.

7.3 Charleroi
7.31. History

The town differs from other cities covered in this book as it is neither ancient nor large, and is the relatively modest but proud commercial and communications hub of what was for over a century one of the densest accretions in Europe of coal-mining, steel-making, and metal-based industries. A chain of related towns lies around Charleroi (like the British 'Black Country' or the Ruhr) and when the Ministry were deciding on future public transport policy a light-rail network appeared to be a reasonable solution, both to address problems of mobility in an increasingly busy area, and also as a means of serving planned new satellite housing schemes which would replace the notably poor legacy of older dwellings. It was on this basis that a massive master-plan was adopted in 1970, with new lines fanning out from a circular pre-metro line around the centre of Charleroi itself.

This would be built partly on viaduct and partly in shallow subway according to varying topographical considerations (the town centre stands on a hill high above the river valley). Other radial lines would include tunnelled sections, and some would make use of former industrial railway routes. Map 27 is a contemporary diagram illustrating the ambitious extent and nature of the routes as originally proposed.

Charleroi had had its own urban metre-gauge tramway system but was also the hub of a vast network of Vicinal lines stretching particularly to the east, north and west (and indeed reaching as far as Valenciennes, Mons, Namur, and Brussels). Despite some post-war extensions this network was already in decline by the time the Charleroi pre-metro plans were adopted, but some of the proposed light rail routes replaced Vicinal lines and at least in an interim stage it was intended that the pre-metro would link to upgraded existing tram lines.

Most of these intentions came to nothing. A massive decline in steel-making and related industries from the 1980s devastated local employment, and resulting population downturn rendered the planned new towns unnecessary.

A parallel programme of motorway construction made traffic circulation easier. The economic argument for a large light rail system diminished, and all the town trams and most of the Vicinal lines closed between 1957 and 1993.

RIGHT: Map 27. The original proposals for an extensive light-rail network around Charleroi envisaged surface, tunnel, and elevated construction, but only a fraction has been built.

The westerly pre-metro line reaches out to Petria, where a short subway section ends at a surface station, with traditional roadside running to Anderlues beyond.

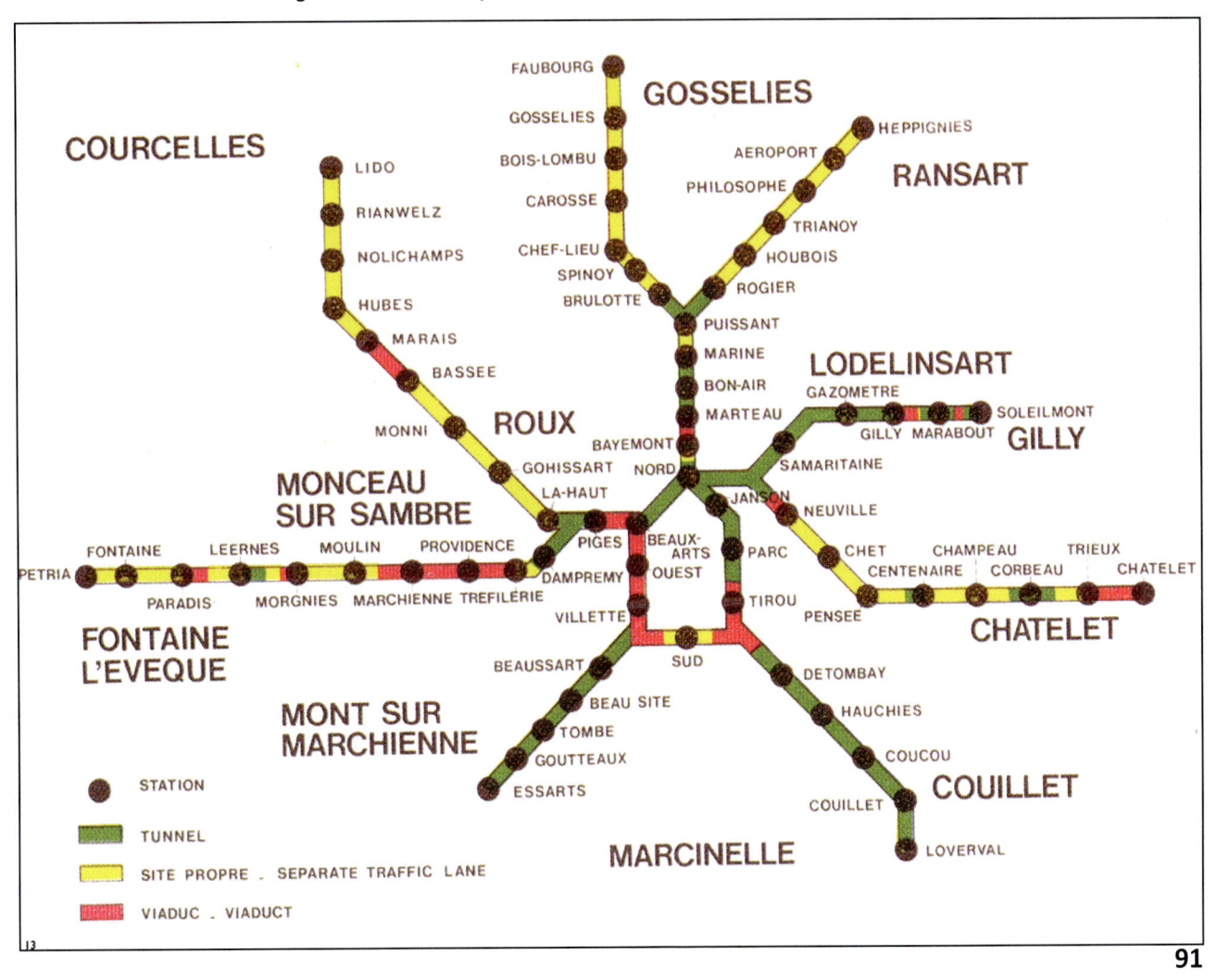

91

7.32 Pre-metro

Nonetheless work proceeded on a core pre-metro system and beginning with a short elevated section in 1976 the Charleroi 'ring' was eventually completed in 2012. A long westerly radial line, almost all on new right-of-way, was completed to Pétria in 1986, its final section to Anderlues retaining a traditional Vicinal roadside routeway by then unique in Belgium. This section is at last being partly rebuilt in 2017-8. Similarly after many delays a radial line to the north-east was completed to Soleilmont, partly in tunnel, in 2012. Another line to Centenaire in the south-east was completed by 1985 but has yet to be opened. A mainly street-based line north to Gosselies opened in 2013.The table shows the sequence of pre-metro opening.

Opening of Charleroi Pre-Metro
sections 1976 – 2012

Section	Date	Stations
Sud - Villette	21 June 1976	Sud, Villette
Villette – Piges	30 June 198	Ouest*, Piges
Morgnies – Paradis	21 June 1982	Paradis, Leernes, Morgnies
Villette – Beaux Arts	24 May 1983	Beaux Arts
Piges – Dampremy	24 May 1983	Dampremy
Paradis – Petria	30 August 1986	Fontaine, Pétria
Dampremy – Morgnies	22 August 1992	Providence, De Cartier*, Moulin
Beaux Arts – Gilly	28 August 1992	Waterloo*, Samaritaine*, Gazomètre*, Gilly*
Waterloo – Parc	30 August 1996	Janson*, Parc*
Sud - Parc	27 February 2012	Tirou
Gilly - Soleimont	27 February 2012	Marabou, Sart-Culpart, Soleimont
(Piges – Gosselies, mostly street running)	21 June 2013	*18 stops*
Waterloo - Centenaire	*Completed 1985 (Unopened)*	*Neuville*, Chet*, Pensée, Centenaire*.*

(*: subway stations)
Source: T2000

After many years delay the town centre circle was completed in 2012, with extension southwards of the existing tunnel and a new street-based link to Gare du Sud. This 2011 view shows the portal of the extended tunnel under construction, at the approach to what later became Tirou station.

The 1992 extension eastwards to Gilly was in tunnel: the terminus station sported the national 'metro' identifier.

The roadbed east of Gilly had been prepared in the 1980s but lay unfinished for thirty years. In 2012 the system was extended through two intermediate surface stations to a terminus at Soleilmont, laid out as a model intermodal and park-and-ride site.

Operating north towards the extended tunnel a BN car leaves Tirou station.

7.33 Rolling stock

Built by BN in Brugge the 55 articulated cars delivered to Charleroi between 1977 and 1985 were an accomplished approach to the 'hybrid' issues which we have considered before. They represent a reasonable compromise between the needs of street-based operation, and efficient service in pre-metro conditions. Double-ended (unlike their cousins on the coastal line) the handsomely-styled cars offer an attractive and practical interior, with generous standing lobbies around the four pairs of double doors on each side, comfortable facing seats in two-and-one configuration, and separate drivers' cabs. Originally intended for crew operation the cars were initially equipped with conductors' positions at each end but these were removed before service began. There are retractable steps, but the body structure prevents level access from high-level platforms. All the BN cars were fitted with multiple-unit couplers, and although these have been removed from the coastal fleet they are retained in Charleroi and are used for some peak-hour workings. The cars have monomotor bogies and chopper control, and are equipped with automatic braking and speed control equipment for pre-metro operation. They are geared for a maximum speed of 65 km/h.

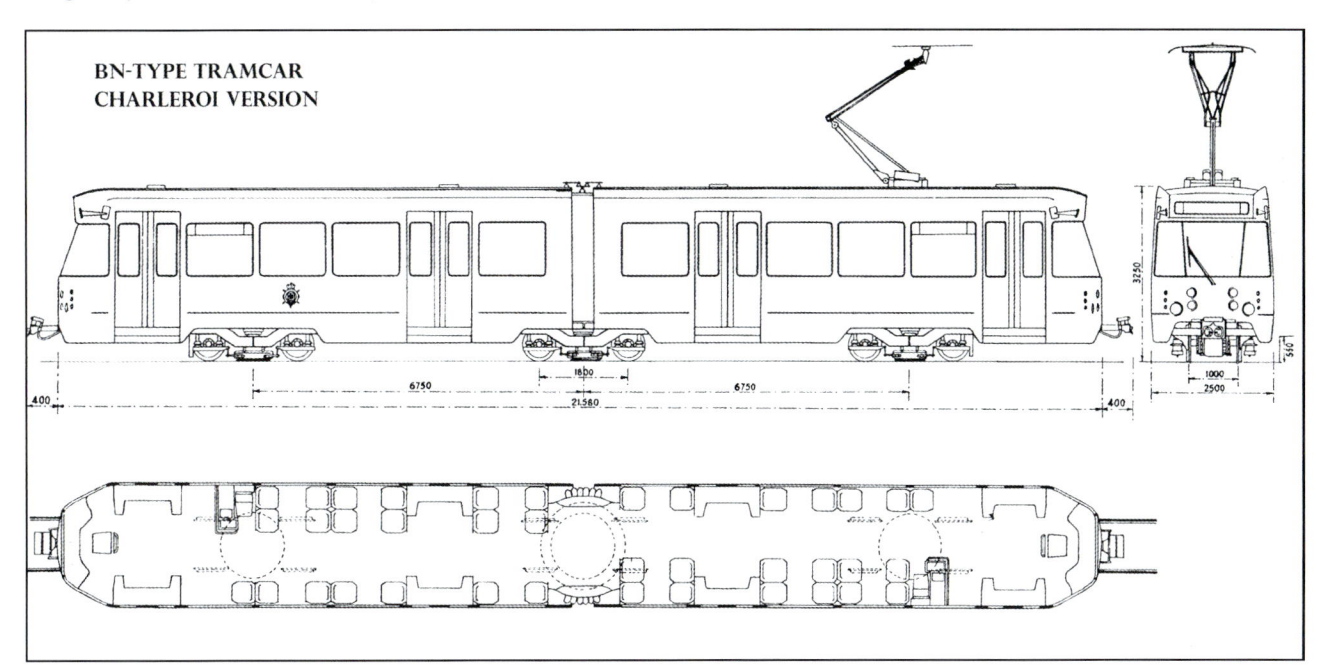

Diagram of a BN car as built, showing the practical and spacious internal layout. The conductors' desks were removed before entry to service.

Details of BN cars

Fleet numbers	Type	Date	Length m	Weight t	Motors kW	Width m	Seats/stand
6100 – 54	BN	1980-82	22.88	32.50	2 x 215	2.50	38/146

Six-axle articulated, double-ended. Original numbering in the Vicinal fleet since replaced by TEC '7400' series numbering. Not all cars now serviceable.

A smart BN car under repair, showing the folding steps and the structural impediment to level boarding. These trams are now being refurbished.

The 49 remaining operational cars meet well the mixed requirements of the unusual Charleroi system and funds were recently allocated for their refurbishment.

So the 29.5km system which now exists in Charleroi is a curious hybrid, a town centre core in impressive pre-metro style, a northerly line laid almost entirely on street, a segregated easterly line partly in tunnel, and a long antenna to the west, mostly either on segregated surface track or elevated but ending as a roadside semi-rural tramway. In a modest way this small network continues to contribute to the commercial vitality of Charleroi, although it falls short of what was originally intended.

A detailed history of Charleroi's tramways, with maps, is contained in our book Charleroi's Trams since 1940 (2013), available from the LRTA.

7.4 Liège

Although historically and topographically different Liège resembles Charleroi as a one-time centre of steel-making and related industries. Its location on the steep banks of the River Meuse makes it a serpentine, constricted city, stretching over a long distance and notable for its congested north-south transport. This is a hilly, gritty place, memorably and accurately described by Jonathan Meades as 'the Yorkshire of Belgium'. Its complex former tramways, together with a network of local Vicinal lines, had all disappeared by 1968, never having been significantly modernised. Inclusion in the Ministry's 1962 programme of provincial light rail projects meant that a new project was under consideration just as the last of the existing lines was facing abandonment. The subsequent story is one of vacillation and disappointment.

The congested streets of the west bank of the river dictated a tunnelled transport system and work began in the early 1970s on difficult excavation in the neighbourhood of the main Guillemins railway station and on the river bank further north. The single route would have stretched the length of the built-up area, and would have been a full metro system, without street-running, although it seems questionable that the volume of traffic would have justified this. The downturn in the economy, and a more sceptical central government, which also bore down on projects in Antwerpen and Brussels, brought work to an end after about 800m of tunnelling had been completed (this now stands idle, used only as storage space). Planning then turned towards another much-vaunted Belgian project of the era, a system called 'TAU' (Transport Automatisé Urbain) which it was hoped would compete successfully with the French 'VAL' system (as used in Lille and elsewhere), opening up export opportunities. This employed much smaller cars, thus reducing excavation costs, and because it was driverless and fully automated would also lessen operational expenses. Liège would have been a demonstration of the system's potential. Substantial central government funds had been made available to develop the project, in association with the (then) Belgian manufacturers BN. The TAU route was to have been about 16km long, with 27 stations between Pont de Seraing in the south and Herstal in the north. It would not, however, have made use of the completed structures built for the previous Metro project.

In turn further financial troubles, the results of regionalisation and hence the end of full central government capital funding, and a clearer understanding of the limitations on export opportunities, brought this initiative to an end in the later 1980s. The new Regional Government subsequently considered the transport problems of the city over several years, and in 2006 produced a third plan, for a 17.5km conventional tramway between Herstal (Esperance) and Jemeppe, with a short branch to Bressoux. This was subsequently reduced to an initial 11km line between Sclessin and Coronmeuse with 21 stops. Although contracts were signed in 2015, and completion was planned within three years, the project has been further delayed, although work is now evident on parts of the abbreviated route with completion now expected by 2022.

Although other projects arising from the Ministry's 1962 initiative developed in ways differing from original conceptions, this was the only one which has so far failed to make much progress.

8. The STIB Metro Develops

We have seen how a full Metro service began in Brussels in 1976, first with the conversion of the east-west line 1, and its extension eastwards initially to Tomberg (1A) and Beaulieu (1B) on the branches diverging at Merode. In the westerly direction the line was extended first to a new underground station at Ste-Catherine (1977) and later to Beekkant (1981).

On the inner ring (line 2) the pre-metro line between Rogier and Louise was converted in 1988, and simultaneously extended westwards to Simonis, and southwards to Gare du Midi.

In this chapter we look at the growth of the Metro, rapid at first, and at its operation today.

The original metro cars never appeared in the newest STIB livery, but the BOA trains have done. These trains are seen on the Herrmann-Debroux branch east of Delta where the line is laid in the centre of the E411 expressway.

A characteristic Metro scene at Arts-Loi on the Petite Ceinture line 2, showing the distinctive design of the unit ends, largely unchanged since the first line opened in 1976. This had been a pre-metro station served by trams until 1988.

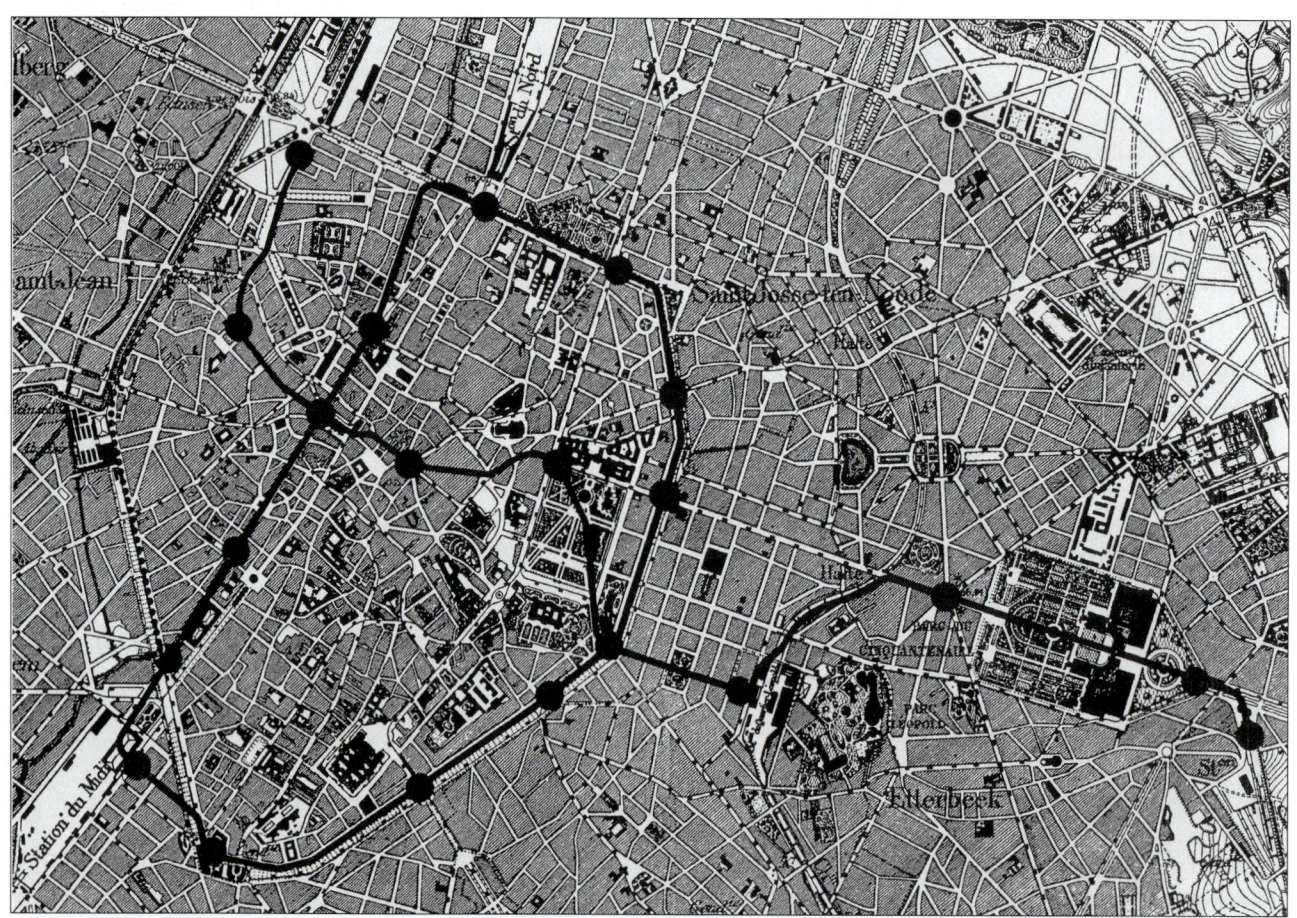

Map 28. The 1926 plan for a conventional metro in Brussels, showing how much it prefigured what was actually achieved after 1969.

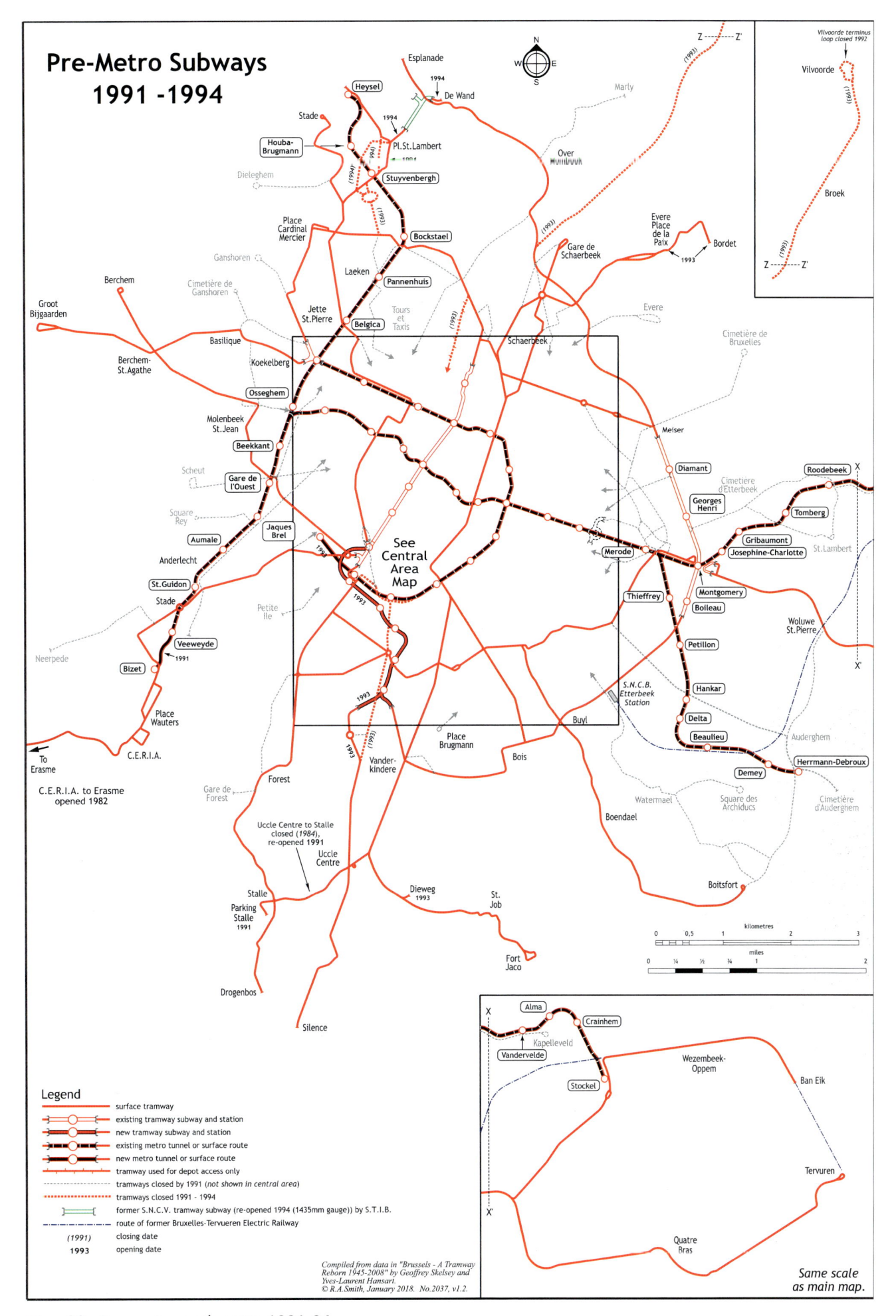

Map 29. Pre-metro and metro 1991-94

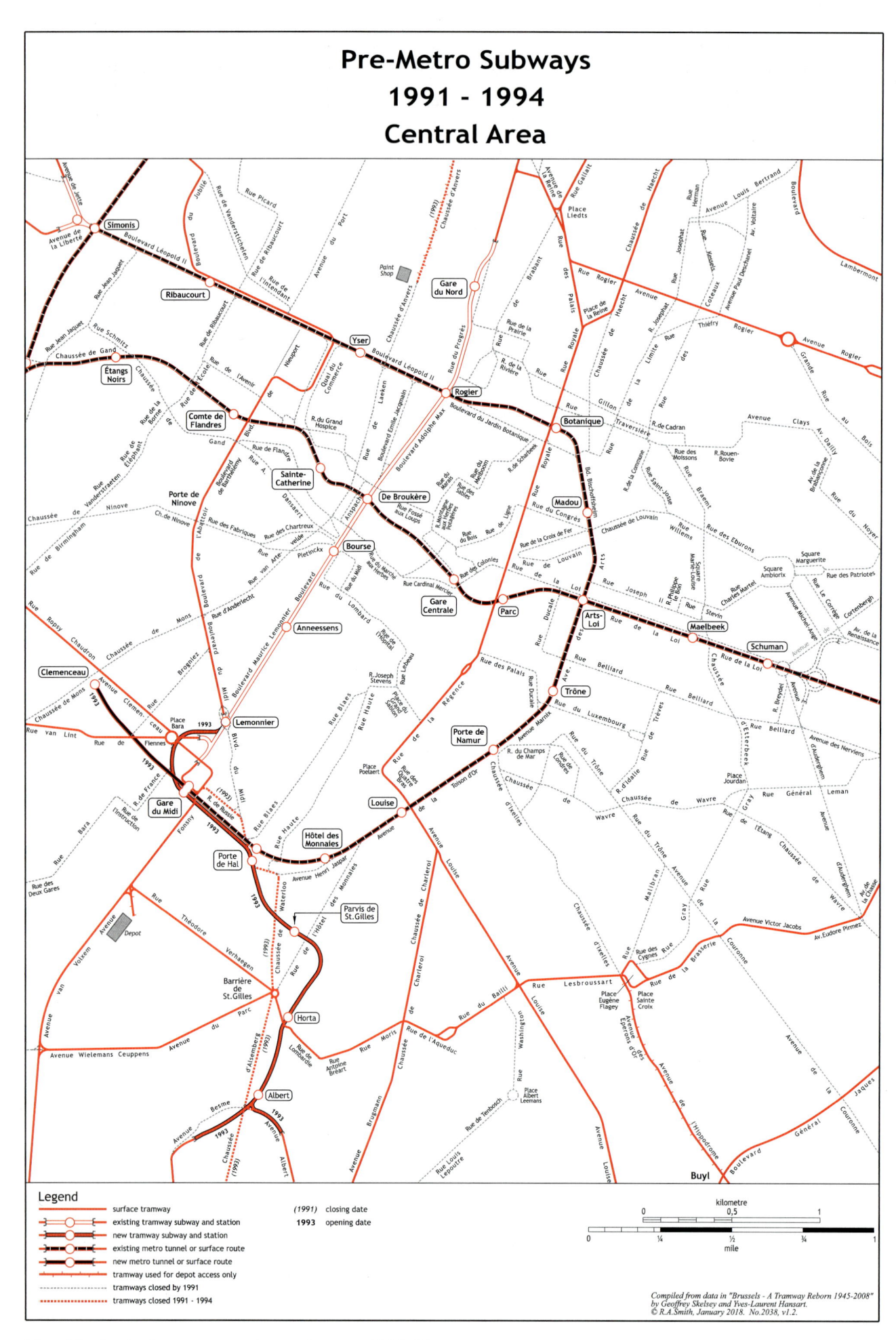

Map 30. Central area pre-metro and metro 1991-94

8. A Brussels Metro in 1926?

It is a pleasant preliminary to find that consideration of underground rail service in Brussels did not originate in the 1950s but long before. Apart from plans for the 'Junction' main line railway connecting the Midi and Nord stations (referred to in detail above) there were successive proposals for a conventional Metro on the Parisian model, starting in 1913. In 1926 plans by a consortium then engaged in improving and electrifying the Brussels –Tervuren Railway (BT) made concrete proposals for a comprehensive 13km central area network (illustrated on map 28) which almost exactly prefigured several of the routes later adopted for the pre-metro. An east-west line and a circular route were envisaged, with a total of 21 closely-spaced stations.

Trains would have consisted of three cars of Parisian style, with overhead electrification at 1500v d.c. Given the promoters' association with the BT project, which eventually terminated at the Gare du Luxembourg, to be served by the new metro, it is easy to envisage these plans together as the basis for what could have been a comprehensive urban and suburban system. Although authorized the difficulty of raising capital, and the onset of the Great Depression, brought the proposals to an end. There were to be no further proposals until after the War.

The following table shows the chronology of actual metro development after 1976.

Chronology of Metro construction 1969 - 2009

8.1 A Designer Metro

It is hard to imagine nowadays, when prestigious and design-conscious metros open nearly every year, just how utilitarian most of them were fifty years ago. In 1976 Brussels was one amongst 47 metro systems in the world, a total which had roughly doubled since 1939 and has more than trebled since. They were then largely a European and (to a lesser extent) North American phenomenon. Whilst there had been early attempts at what would now be called 'styling' – Guimard's exquisite stairways and pavilions in Paris, Green's more robust ox-blood stations and decorative tiles in London—these had been overlain with grime and alterations, and the predominant impression was of systemic neglect. New York was then near its nadir (and is in crisis again in 2018).

Even the earlier postwar Metros -- Stockholm (1950), Toronto (1954) for example-- were clean and functional but then had a sort of bland uniformity, akin to a budget hotel. A change perhaps began with Montréal (1966) where a conscious attempt to reproduce the fabled style and élan of France delivered striking artistic surroundings and highly distinctive trains and stations. Such ventures required generous supplementary funding, although in the context of construction costs the embellishments were relatively cheap. But in Anglo-Saxon eyes at least they would have been regarded as avoidable frivolities: the furore which followed the commissioning of Epstein's sculptures at 55 Broadway, London Underground's headquarters, took generations to live down.

Line 1 (east-west)
Original section, Ste-Catherine – Schuman, opened as Pre-Metro on 17 December 1969, Metro on 20 September 1976, extended to Beaulieu and Tomberg (total line length 9.03 km).

Extended eastwards
Beaulieu - Demey (17 June 1977)
Tomberg - Alma (7 May 1982)
Demey - Herrmann-Debroux (23 March 1985)
Alma – Stockel (31 August 1988)

Extended westwards
De Brouckere – Ste-Catherine (13 April 1977)
Ste-Catherine – Beekkant (8 May 1981)
Beekkant – Bockstael (6 October 1982)
Beekkant – Saint-Guidon (6 October 1982)
Bockstael – Heysel (surface station) (10 May 1985)
Saint-Guidon – Veeweyde (5 July 1985)
Veeweyde – Bizet (10 January 1992)
Bizet – Erasme (15 September 2003)
Heysel (new station) – Roi Baudouin (25 August 1998)

Line 2 (Inner Ring)
Original section, Rogier – Louise, opened as Pre-Metro between 1970 and 1985, as Metro on 2 October 1988 to Gare du Midi and Simonis (total line length 6.67 km).

Extended westwards
Gare du Midi – Clemenceau (18 June 1993)
Clemenceau – Delacroix (4 September 2006)
Delacroix – Gare de l'Ouest (4 April 2009)

In preparation for the launch of the Metro in 1976 a mock-up of a metro car was built and displayed to the public at various sites, including (improbably) here at Woluwe Depot. The picture also shows the relative size of a 5000-type tram (on the left) and the new cars: the former was serving as a temporary post office. [MJR]

Brussels followed the 'artistic' trend, but more fully over time and more extensively. The Metro would not only be a practical transport system but its trains and stations would make a statement, about the aspirations and culture of the nation and the city, but also the vitality of a confident transport system. Although there were recurrent elements, such as signage and the station identifier, as well as the trains of course, there were no standard design elements or colours in the stations, each of which was a creation complete in itself. Brussels stations never approached the lavish surroundings of the 1935 Moskva Metro, and its successors in other cities of the former USSR, but the object –to demonstrate national pride and achievement- was perhaps similar.

As it became clear in the late sixties that the transition to Metro would take place earlier rather than in the indefinite future the STIB commissioned the design firm of Neerman & Cie to advise on the rolling stock. Eighteen months later the Ministry of Communications established a design group embodying both the STIB and their consultants but also representatives of government bodies and the principal Belgian manufacturers.

Underlying this strategy, as with other initiatives of the period, was the hope that Belgium could recapture its international preeminence in rolling stock design and manufacture, and establish a lucrative export trade. The Belgian state has, ever since its foundation, taken an active role in encouraging industry and promoting exports, and underlying the attempts to devise an impressive and effective metro system was the hope of an opportunity to showcase the nation's technical achievements. Apart from a modest success in Manila these efforts were unavailing. A sample Metro car was produced for a major exhibition at the Design Centre in the Galerie Ravenstein in 1972, and although the eventual production vehicles differed in some respects, the mock-up and models embodied many of the eventual elements which have lasted remarkably well. Work also took place on standards for metro stations, including signage.

An early view of a metro car under construction: note the massive coupler, but also the fact that (unlike many American and British cars of the era) there were no end doors).

A highly individual 'Metro' identifier was adopted from the start, and is still used in Brussels.

The distinctive livery of the original Metro cars shines brightly as a two-car train leaves Heysel for the centre of the city, visible in the far distance. The tramway is on the left. [Courtesy LRTA Collection/Online Transport Archive]

Map 31. The Metro line diagram from 2000 presents a slightly confusing impression, at least to inexperienced travellers. The 'full' Metro system is shown comprehensively, including the two then-unfinished sections. But the pre-metro routes are confined to the tunnelled sections, with no indication of extent beyond the respective portals apart from a service list below. More recent versions have improved on this, but illustrate a somewhat tentative approach. Strangers would be baffled by, for instance, the inclusion of the Poincaré and Jamar ramps. What the diagram really reveals is the random set of services then using the north-south axis.

101

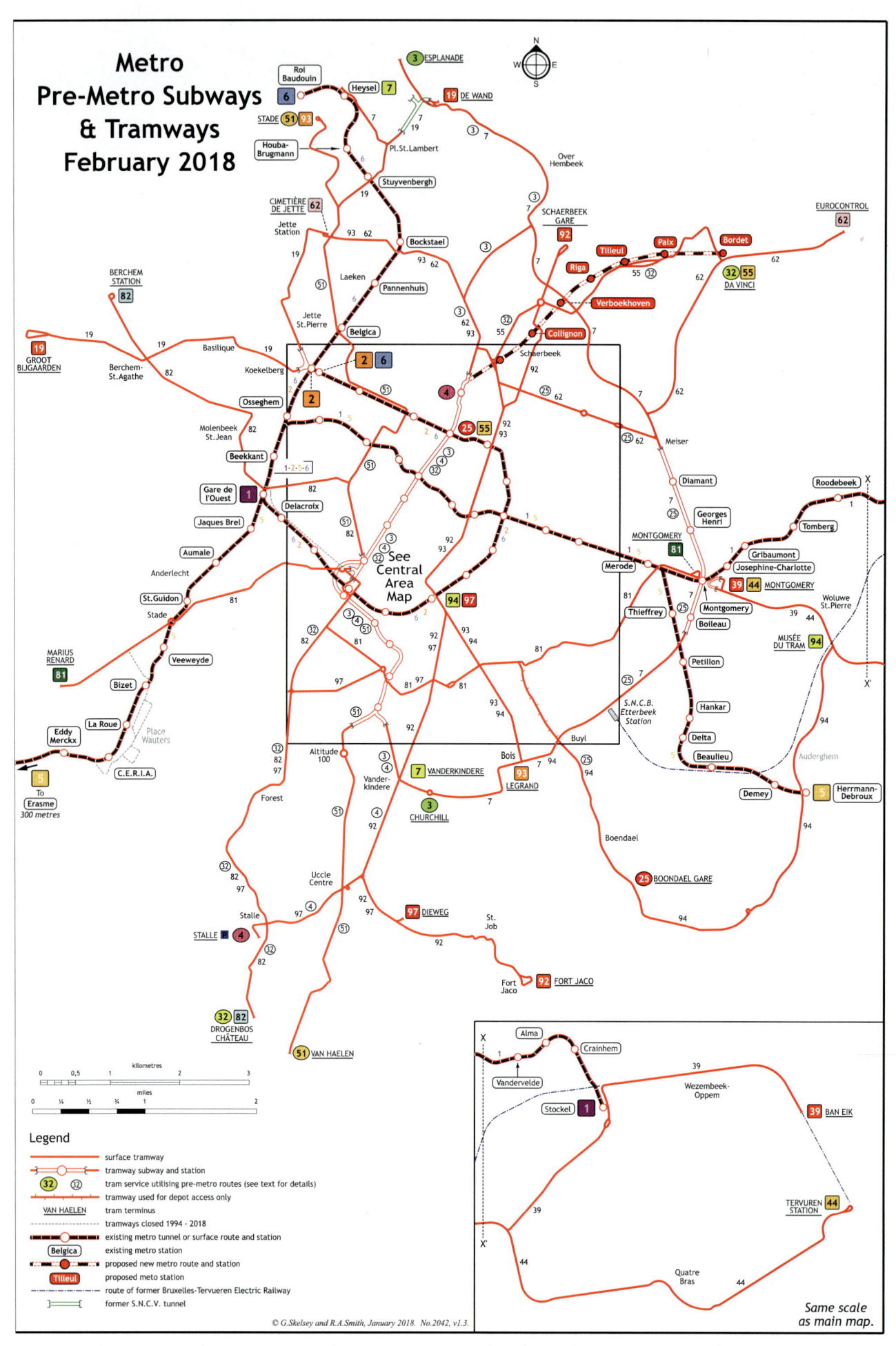

Map 32. The Metro and pre-metro in February 2018. Note the planned extension towards Bordet which will complement the converted line 3. [RAS]

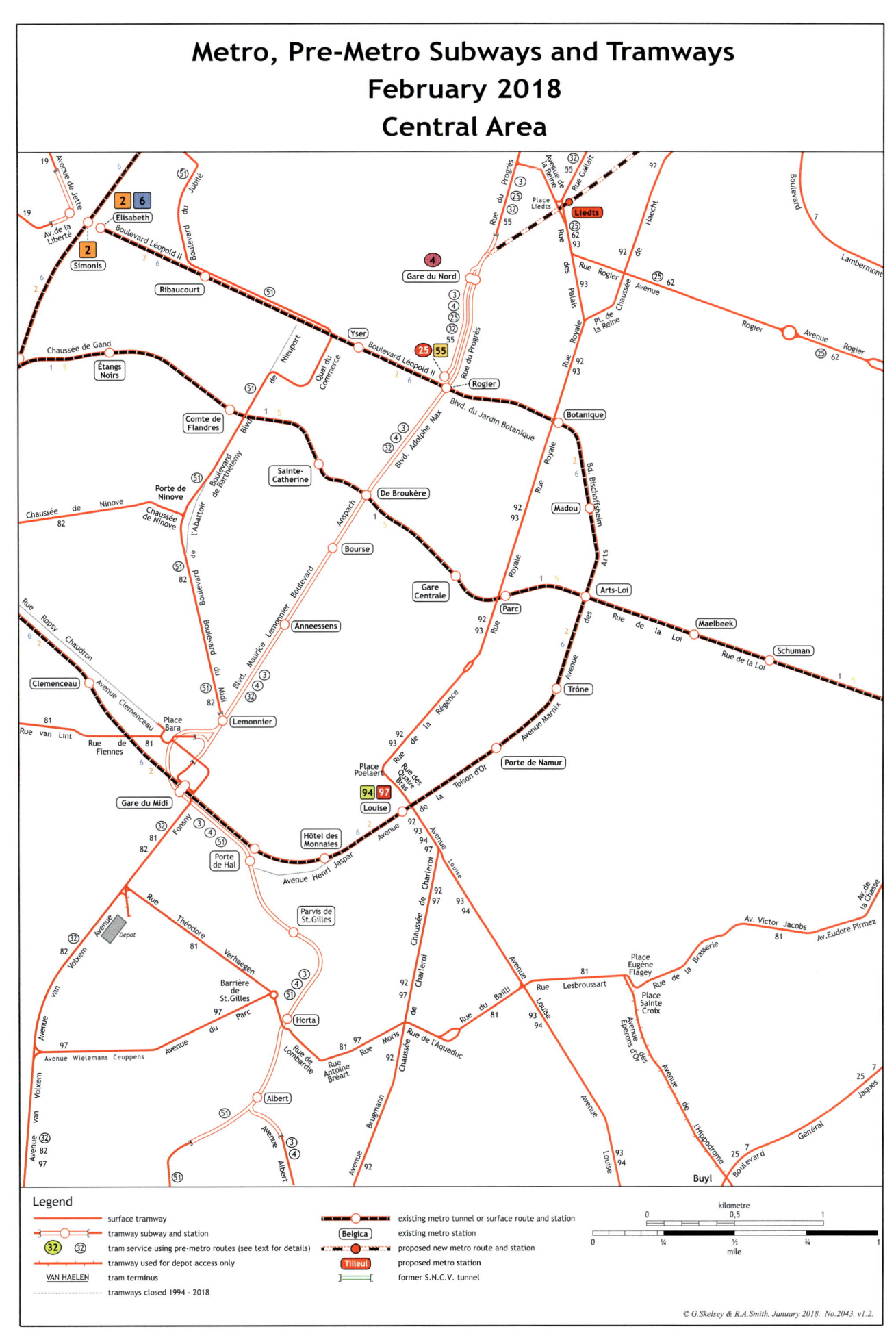

Metro, Pre-Metro Subways and Tramways
February 2018
Central Area

Map 33. Inner area metro and pre-metro, February 2018. Note (upper centre) the duplicate tracks between Gare du Nord and the terminal bays at Rogier which will accommodate the feeder tramways from the north after about 2028 [RAS]

103

8.2 Trains

The original design eventually amounted to 217 cars, delivered in five batches between 1974 and 1999, and with the successive builds largely indistinguishable. Units were originally formed of two cars, but increasing traffic soon led to the construction of non-driving motor cars for formation into some three-car sets. There are now 53 two-car and 37 three-car sets, normally forming four- or five-car trains, the latter being the limit for existing platforms.

The new cars were 2.70m wide (the trams were then 2.20m wide), 18.2m long, and in two-car format accommodated 420 passengers of whom 80 were seated. For Metro operation the line voltage was raised from 700 to 900 V d.c., and additional substation and supply capacity was needed to meet the higher power consumption of the trains. Consideration was given to retaining overhead line equipment, but in the event under-running third-rail supply was adopted (thus improving reliability in snow or ice). Performance of the trains was impressive, and is still noticeably so today: every axle is powered, with one motor in each bogie rated at 266 kWh. Acceleration is at 1.16m/s from 0 to 30km/h, 0.90m/s from 30 to 60km/h. 72km/h could be achieved in normal operation. Successive deliveries of cars retained most of the original features. The distinctive appearance of the cars has dominated the system for over forty years.

Most notable is the orange/red vehicle end, slightly bulbous and with a single driving window. The sides of the cars are in unpainted alloy with an orange stripe at floor level. Four pairs of double doors provide easy and rapid access: they are in most cases opened by the passenger after release by the driver, who closes them. Interiors are also distinctive. Most seats are in groups of four with headrests and grab handles for standing passengers; to create more standing space some seats have more recently been removed from bays at the outer ends of the units. There are audible interior announcements as well as visual displays, in both languages.

An order for a wholly new design of thirty 'BOA' metro cars in six-car units with wide inter-car gangways was placed in 2005. These add much-needed capacity on the crowded east-west line. On the new BOA trains seats are longitudinal, leading to reduced seating capacity but far more standing space. They appear in a grey livery with highlighted doors.

The STIB intend to introduce fully automatic operation on the east-west route in order to improve regularity and performance. Given the age of the original cars, fleet replacement, with fully automatic driving capability, is a coming issue, and 43 new generation trains were ordered from CAF in 2017, with more to come. A full-size mockup of the new fleet, to be designated 'M7', was displayed in January 2018.

8.3 Routes

In 1966, as ideas for the gradual creation of a city-wide metro system emerged, the aspirations were bold and the expected realisation rapid. The following table shows how the network was originally envisaged.

Schedule of Metro Construction as planned in 1966 (*Built as pre-metro)

Line No	Itinerary	Completion dates
1	Berchem/Gare de l'Ouest – Etangs Noirs – Ste-Catherine* – Schuman* – Montgomery –Square Leopold II/Pont de Woluwe.	1969-87
2	Rond Point du Meir – Gare du Midi – Louise* - Rogier* - Simonis – Place Bockstael – Centenaire.	1969-79
3	Uccle (Place Danco) – Albert* - Gare du Midi* - Gare du Nord* - Liedts – Gare de Schaerbeek – Esplanade.	1975-82
4	Bois/Boondael - Place Flagey – Parc – Botanique – Ste-Marie - Liedts/Meiser (not built).	1984-87

This would have produced a network of 57.6 route/km, mostly underground (the actual 2017 network, including pre-metro, totals 52km with 69 stations). Completion was optimistically expected in 1987. There would have been 95 stations, at an average spacing of 630 metres, and estimated rolling stock requirements were 204 units (408 cars). This would have given a

The final stages of conversion of pre-metro to metro included alterations to the lowered platform sections in anticipation of hectic work during the changeover period. At the interchange station at Arts-Loi on line 1 the third rail had already been laid on 19 April 1976. [MJR]

relative route length of 44 km per million inhabitants, considerably in excess of the corresponding figure for Paris. Although the aspirations for a Metro network changed over time elements of the present system can be identified from the original goals set in the late 1960s. There were however important differences.

At the western end of Line 1 the initial idea was for extension further westwards under the Chaussée de Gand to terminate at Berchem (provision was made when the Heysel line was built to allow this in future). There would also have been a branch alongside the railway to Gare de l'Ouest. Line 2 was not envisaged as a circle, but its northern sector would also have been extended westwards, turning north-west alongside the railway on the general routing to Heysel later adopted for Line 1. The southern sector of line 2 would have extended westwards from Midi to terminate at Rond Point du Meir. The coverage of these proposals was largely achieved by the lines as built, but their linkages to the centre were different. A glance at the present Metro plan will show how the network is somewhat unbalanced and the original intentions would have addressed this.

Line number 4 was reserved for an additional north-west – south-east cross-city line, which has had various routings over the planning period, but in the 1960s was envisaged as running between Boondael and Place Liedts via Flagey, Porte de Namur, Parc, and Botanique. Some constructional work was undertaken at the intersections with other lines in anticipation of the new routing.

Notably missing from the list is line 5 (the eastern peripheral line) which was to have formed a long circumferential line stretching around much of the city between Gare du Nord and Gare du Midi, incorporating the existing section of pre-metro south of Meiser. This is unlikely to come to fruition in the foreseeable future.

The core of the present Metro is the east-west and circle elements which were converted from pre-metro between 1976 and 1988. Following the initial conversion and extension of the east-west line a remarkable period of expansion followed, with regular extensions between 1977 and 1991 in a period of relatively generous central government funding, and a pro-metro policy generally.

The initial operation by single-ended tramcars naturally imposed right-hand running on the system, which continued on conversion to Metro (at variance with the practice on the Belgian main-line railways). There were however later exceptions, as mentioned below. The profile of the underground lines involved some notable ascents, for example eastbound between De Brouckère and Parc, and northbound between Porte de Hal and Porte de Namur, and this required high installed power in the trains.

Briefly to describe the creation of the present Metro system, the earliest extensions expanded the east-west line (1). The southern branch from Merode was largely on surface, partly alongside the SNCB

The under-running third rail is largely immune to the effects of ice and snow: a wintry scene with a BOA train on the Herrmann-Debroux branch.

When pre-metro operation of line 1 was envisaged as continuing indefinitely the intention was to upgrade the long light-rail lines east of Montgomery as part of the upgraded network. Later there were thoughts that the metro itself would extend to Tervuren, but traffic analysis showed this to be unjustified and the route remains a high-quality feeder tramway. This is Avenue de Tervuren on a chilly day in February 2001.

The Heysel branch passes close to the Atomium, and there is an interchange with the adjoining tramway. As explained in the text, on this section metro trains run on the left.

The new Erasme terminus adopted a tent-like structure.

peripheral line. A short extension from Beaulieu to Demey (840 m) was opened in 1977, with tracks laid on the centre strip of an expressway. A further underground extension to a terminus at Herrmann-Debroux (570 m) opened in 1985, where from 2006 there was connection with a new feeder tramway.

On the northerly branch from Merode the line was completely in tunnel, extending to Alma (2.17 km) in 1982 and Stockel (1.28 km) in 1988. At the latter station a fine new express tramway linked the metro to the affluent north-eastern suburbs, terminating at Ban Eik. This was laid on the disused roadbed of the former Brussels-Tervuren railway, and is a model of modern light-rail transit.

West of De Brouckère the line was extended a short distance to a new underground station at Ste-Catherine in 1977 (420 m), but beyond there the next stage took several years to complete and involved engineering work as demanding as any on the system. The 1.78km line towards Beekkant crossed low-lying ground riddled with old watercourses, as well as the large and busy Charleroi Canal. The future Comte de Flandres station was near the western canal bank, and the location demanded that the Metro passed under the canal, which had to be kept open for traffic. Work started at the site on 26th April 1975 and continued until 30th April 1979, requiring property demolition, works to ensure continued passage on the canal, and a vast excavation over 20 metres deep to accommodate the station. Chemical soil consolidation and continuous pumping were necessary. As finished in May 1981 this station is perhaps the most impressive on the network, with a vast platform hall decorated with enigmatic sculptures suspended from the high ceiling. The line was eventually opened on 8 May 1981. At Beekkant the layout dictated a curious operating arrangement which survived for almost thirty years. The ensuing extensions were laid in part alongside the SNCB West Side peripheral line, and the metro tracks from Ste-Catherine trailed into this at the four-platform Beekkant station. Metro trains on line 1A proceeding towards Heysel had therefore to reverse there, and to facilitate non-conflicting operation in the station the running direction to the north was on the left.

Three car sets were introduced to cater for increasing traffic, with an intermediate non-driving motor. This is Porte de Namur in 2001.

The long tram extension from CERIA to Erasme, opened in 1982, was designed as a surface premetro, with the highest standards of catenary overhead and track. In the event the extension of the metro from Veeweyde and Bizet, completed in 2003, subsumed this section which meant that it was relatively short lived as a tramway. [MJR]

The expansion of the Metro was marked by several interim terminals whilst construction proceeded, but this one lasted for years. We are looking outbound at Heysel in 1985, with the simple surface station on the left and the altered tram terminus on the right, although at this time the tram service ran only for special events. The Metro was further extended in 1998. On the far left the overhead lines just visible were for a demonstration of the 'GLT' guided transit system, staged for the UITP. [Courtesy LRTA Collection/Online Transport Archive]

Metro style on the Fête Nationale, 21 July 2017

107

The line was further extended to Bockstael (3.37km) in 1982 and onwards to a simple surface station at Heysel (2.29km) in 1985, adjoining the elaborate tram terminus. Southwards from Beekkant the line reached St-Guidon (1982, 3km), Veeweyde (1985, 720m), Bizet (1991, 600m), and onwards to its final terminus at Erasme in 2003 (2.78km), adjoining a major teaching hospital, with intermediate stations at La Roue, CERIA, and Eddy Merckx (named after the celebrated Belgian cyclist). The final section of the new line was built on the route of a former surface tram extension. Pressures on existing storage and maintenance space, as well as the needs of a line 3 Metro, prompted a decision in 2016 to build a third Metro depot here at Erasme. A new underground Heysel station was opened in 1998, and the line extended a short distance to terminate at Roi Baudouin (505m). Until the total reordering of metro services in 2009 trains on the Heysel branch worked to and from Herrmann-Debroux (line 1A) and those on the Erasme branch were linked to Stockel (line 1B).

The completion of the inner ring (then line 2) was a protracted task, partly occasioned by the complexity of construction work and the commissioning of a new connection at Gare de l'Ouest with the existing line 1. A new depot was also required. The line reached Clemenceau (500m) in 1993, but as had been the case at Comte de Flandres twenty years earlier (see page 106) the crossing of the Charleroi Canal again caused constructional headaches. Here the line passes over the canal near Delacroix station, and building took several years to complete. The line was opened to Delacroix (650m) in 2006 but the final link at Gare de l'Ouest (1.15km) was not finished until 2009. The Delacroix extension enabled abandonment in 2007 of the parallel tramway, along Avenue Clemenceau, with its badly deteriorated tracks. (A reopened route along Chaussée de Ninove allowed continued tram service along the corridor).

The whole line 2 was ceremonially opened by H.M. King Albert II in a special train on 2nd April 2009. Because the tracks into Gare du Midi rolled over one another to provide cross-platform interchange in the contrary direction the direction of running between there and Beekkant was on the left. This extension completed the circular route originally begun over thirty years earlier and brought underground construction to an end for the time being, after almost continuous work for forty years. With stations placed typically at about 500m intervals the system shows its tramway and pre-metro origins, notably different from 'rapid transit' installations like London's.

With some stations now nearing their half-century a programme of rehabilitation and refurbishment is proceeding, and a noticeable development has been the installation of ticket gates to prevent fraud, and a requirement to check-in and check-out on each journey. Stations are gradually being made more accessible by installation of lifts. New signage in lower case was prepared by the typographer Eric de Berranger in 2006 and has since been extended across the system. Late in 2017 a public consultation opened on future design standards. Wi Fi connection is being provided in all Metro and pre-metro stations, a programme now nearing completion.

8.4 Present service pattern
The linking of the former two lines in 2009 made possible the integration of services, and the pattern became a 'circular' (but not continuous) line 2 (Simonis [Elisabeth, low level] -Beekkant – Simonis [Léopold II, high level] and return); a through service between Erasme and Herrmann-Debroux (line 5); and a new service operating Simonis – Inner Ring – Beekkant – Roi Baudouin, taking in both former lines (line 6). Former line 1 now runs between Stockel and Gare de l'Ouest only. This change opened up through journey possibilities to inner ring stations for some passengers, but involved an additional change for others. The new services were introduced with free travel on 4th April 2009. The changes brought the total Metro route length to 39.9km, with 59 stations (see map 32).

8.5 The Future
Planning and executing the line 3 conversion, together with its northern extension, will occupy the STIB's funds and resources for years to come (this major venture is fully described in the next section). In addition, it is intended to phase in automatic driving, together with new rolling stock, over the next few years. Thereafter extension plans are still speculative, but the Region and the STIB have announced possible future initiatives aimed at relieving serious overcrowding, expected to worsen, at the eastern end of line 1. The two branches diverging at Merode necessarily limit the service which can be provided over each of them, and a solution has been devised: this would require 'unbundling' line 2 and abandoning the 'circle' service. Instead a new line would be constructed from Porte de Namur through Luxembourg (with an SNCB interchange) to Schuman and Merode, where it would take over the Stockel branch. On the northern sector of the Petite Ceinture a new line would diverge at Arts Loi, initially to Luxembourg and later to Vanderkinderere, which would become an intermodal interchange for the southern suburbs. Like all such initiatives there would be winners and losers: some passengers would benefit from improved connectivity, access to the 'European Quarter' would be much improved, and frequencies on the Stockel and Herrmann-Debroux branches could be doubled. But the section of line between Porte de Namur and Arts-Loi would be abandoned, and Trône station closed.

The result would be three wholly separate metro lines, without running junctions. Apart from increased capacity and reliability the new layout would provide for the first time direct connections between Midi and the European Quarters at Schuman and Luxembourg, avoiding the overcrowded interchange at Arts-Loi. If built a Vanderkindere extension would have implications for the future of the Chaussée de Charleroi tramway.

9. Line 3: Twenty-first century questions

The 1993 extension of line 3 ended in twin portals south of Albert. This is the 'Jupiter' ramp, used by cross-city service 51 between Heysel and Van Haelen, most of which is street running.

> As we have seen the original intention was that north-south line 3 would eventually run, as a Metro, between Uccle (Globe) and Albert via Bourse to Schaerbeek (possibly extended to Esplanade). The northern and southern prolongations were not built, but the core section between Nord and Albert was completed as pre-metro in 1993. Despite improvements, problems on the route persisted and by the 2010s further change was planned.

The radical overhaul of line 3 early in the new century transformed the capacity and reliability of the service. In particular, simplification of the service pattern, higher frequency, and the introduction of high-capacity 4000-type cars (each with about 260 places), brought standards approaching those of Metro operation. However problems persisted: the improved service itself created extra traffic, and growing employment along the route drew in more passengers, including those interchanging at Midi, Nord, and De Broukère. Furthermore interchange arrangements with the feeder services at RP Winston Churchill were never wholly satisfactory, especially in poor weather. Within a few years overcrowding was again becoming a problem.

The problems in fact centred less on the core service between Nord and Albert, but on the reliability of journeys as experienced by those feeding into line 3 at Nord, Midi, and Albert, for instance. Here, it was said, persistent road traffic congestion, especially to the north of the centre around what had long been persistent pinch-points such as Place Liedts, was a continuing difficulty. Peak time average speeds here of both trams and buses had fallen below 11 km/h. Such problems impacted adversely on tram service 55 (now a Da Vinci – Rogier shuttle) in particular. For these travellers the line 3 shuttle did not address the needs of the communes north-east of Gare du Nord, which included densely-populated areas, some with social deprivation which would benefit from better links with employment opportunities. Office development was also spreading north-eastwards to take advantage of cheaper property.

The never-rectified working in the original 1957 Constitution Tunnels continued to affect the line. Service 51, which shared the southern part of line 3 which it joined over the junction in the 1957 Midi subway at Poincaré, was also subject to congestion.

At one stage remedial work envisaged at Liedts had included a tram underpass, but this was hard to achieve in practical terms, given the street layout. There was then a proposal for subway extension from Nord as far as the 'Cage aux Ours' (Place Verboekhoven), and the familiar question arose –aired ever since 1969—whether an intermediate pre-metro phase was appropriate or whether the line should be built for Metro operation from the start, as had been the final parts of line 2 from 1988. Given relative operating costs, and the alleged pressures on the core route, policy moved towards a Metro project, a trend consistent with STIB thinking since the 1970s but without a determined contrary voice.

South of Albert the position was less clear-cut, although traffic congestion was also present in Uccle, where north-south arteries were narrow and segregated track hard to achieve. Chaussée d'Alsemberg was another black spot in terms of delays. However the cost of extension southwards was high, and because of the pattern and density of settlement, rewards possibly less. In late 2016 the possibility of extension to St Job, Observatoire, or Uccle-Calevoet was under consideration.

A decision on the core section was announced by the Capital Region Government on 26th March 2015. Line 3 was to be converted to automated metro operation by 2024, with a 5km north-easterly extension diverging from the present route at Gare du Nord towards Bordet. Construction was to begin in 2019,

On the through north-south line at Rogier the interim low platforms have been in use for over forty years but will need reconstruction for Metro use.

The terminal bays at Rogier station were planned in the 1970s as a future terminus for feeder services from the north, connecting into a line 3 Metro. They are now expected to fulfil that purpose.

using a boring machine to lessen surface disturbance. Six closely-spaced intermediate stations were planned, at Place Liedts, Place Collignon, Place Verboekhoven, Square François Riga, Rue du Tilleul, and Place de la Paix, mainly built by a combination of cut-and-cover and enlargement of bored tunnels (see map 33). The technique meant that the line did not have to be built below a main highway and it will in fact be partly located some distance away, which necessarily makes interchange with some existing surface services less convenient. The northern terminus will be at Bordet SNCB station, with a surface extension beyond curving back westwards to a depot within the STIB's existing Haren complex. All stations would be fully accessible.

The existing section of route would be adapted, much as previous pre-metro conversions had been, but major works would be needed at Gare du Midi to by-pass the 1957 subways and to connect with the present deep level tunnels below the railway station. Details of the staging of the work are not yet known: consultation documents make it clear that work would be planned to permit continuous operation of existing services. At the time of publication details remained to be settled after further consideration of environmental and other issues, and full assessment of the costs: a public inquiry into the redevelopment of the Constitution area is expected to begin in June 2018. Albert station was built with future tram termini in mind and incorporates unused additional levels to accommodate trams on (present) services 7 –extended from Vanderkindere- and 51 (south). It would still require major expansion, to provide both a metro terminus and provision for the feeder trams, and the alterations here also are subject to public inquiry. As the tram line south of Albert along Chaussée d'Alsemberg would otherwise be severed from the rest of the system, a new connection was installed in 2017 at the crossing of the Rue de Stalle in Uccle.

During 2018 the aspiration to complete the conversion was moving further into the future, in the light of new regulations, delays arising from the public enquiries, and growing public opposition to surface works on the Bordet extension, notably around Square Riga. 2028 now seemed to be a more realistic objective.

The likely effect on the tram network is discussed on page 104. This will be the greatest disruption to the entire network since conversion of line 2 in 1988.

Schedule speed on the whole line would be 30km/h, and there would be a peak service of twenty trains an hour.

Critique

A decision has been taken and we should of course be glad at such massive investment in public transport along the route. But there is another point of view and it is appropriate to summarise contrary arguments which question whether this conversion represents best value for money. Let's remind ourselves of one inescapable disbenefit: at present passengers to

The 1993 section of line 3 at present incorporates the familiar features of 'temporarily' lowered platform sections for tramway use. At Porte de Hal two generations of trams pass, on the left a long 4000-type car on southbound trunk service 4, and on the right a six-axle car on a short working of northbound service 51, which uses this section of the subway before branching off on street track at Midi, towards Jette.

The old wye-terminus of former service 55 at Silence was removed in 2008 when the line was extended to Van Haelen, but many of the streets south of the subway are narrow. Replacement service 51 will become a shuttle into the Metro at Albert.

North of Gare du Nord service 55 is now a feeder from the important north-eastern suburbs: this is the former Bordet terminus opened in 1993 and now extended to Da Vinci. Although there is segregated track on the extension much of the route is in relatively narrow streets.

The old 52 service, later service 4, already had a largely congestion-free route on its northern sector, improved by a new line along Quai des Usines opened in 2006. This is the junction at Square J. De Trooz, with the new line diverging in the background The view also shows an increasing trend for reserved sections to be shared with buses. In the far distance the imposing 'Royal' Church at Laeken was under repair.

and from five suburban destinations enjoy through services to one part or another of line three; in future only one corridor (part of present tram service 55) will have that advantage, and the majority of present travellers will have to change, some more than once. Notably, the almost wholly segregated Esplanade 'express' line will be relegated to the status of a feeder, its inbound passengers compelled to jostle with those of the other feeders to board already busy metro trains at Nord. Somewhat faster journeys along the core section will be offset by the greater time and inconvenience involved in changing.

But could a continuing but enhanced surface tram operation meet longer-term needs? With certain, cheaper, changes the answer is 'yes'. In particular, diversion around the Constitution tunnels and improvements along the 55-route axis could mean that through service to and from several destinations could be preserved, whilst improved operation of the 4000-type trams --or a more recent successor-- could offer both capacity enhancement and higher scheduled speed at a fraction of the cost. Amongst others Vincent Carton, Professor of Urbanism at EFP Bruxelles, has advanced these and similar arguments, and claims that the present capacity of the core route could be doubled, with an increase in schedule speed from 14 to 18.2 km/h.

There is at least a question to ponder.

On the Vanderkinderere branch south of Albert fine reserved tracks lead into the suburbs, like this line at Longchamp. This will become a high-quality tramway feeder.

Opened in 1991 the large park-and-ride site at 'Parking Stalle' has become an important tramway hub and will continue to be served by a metro feeder.

10. The Brussels Tramways in 2017

We finished our comprehensive Brussels tramway history in 2008 on a note of optimism, as the system had by then decisively thrown off the neglect and disfavour which had blighted it for so long. But we could not have foreseen the surge in traffic and the extent of re-equipment and extension which came in the following years. In this chapter we supplement our 2008 volume and bring the tramway story up to 2017, together with a look ahead at the medium-term future after the conversion in the 2020s of line 3 to Metro.

The southern end of the 93 service to Place Brugmann outlasted most of the minor STIB lines but eventually closed in 1985. In April 1984 an inbound PCC passes in front of the Luxembourg station, in an area then rather run-down but now a centre of the European Quarter.

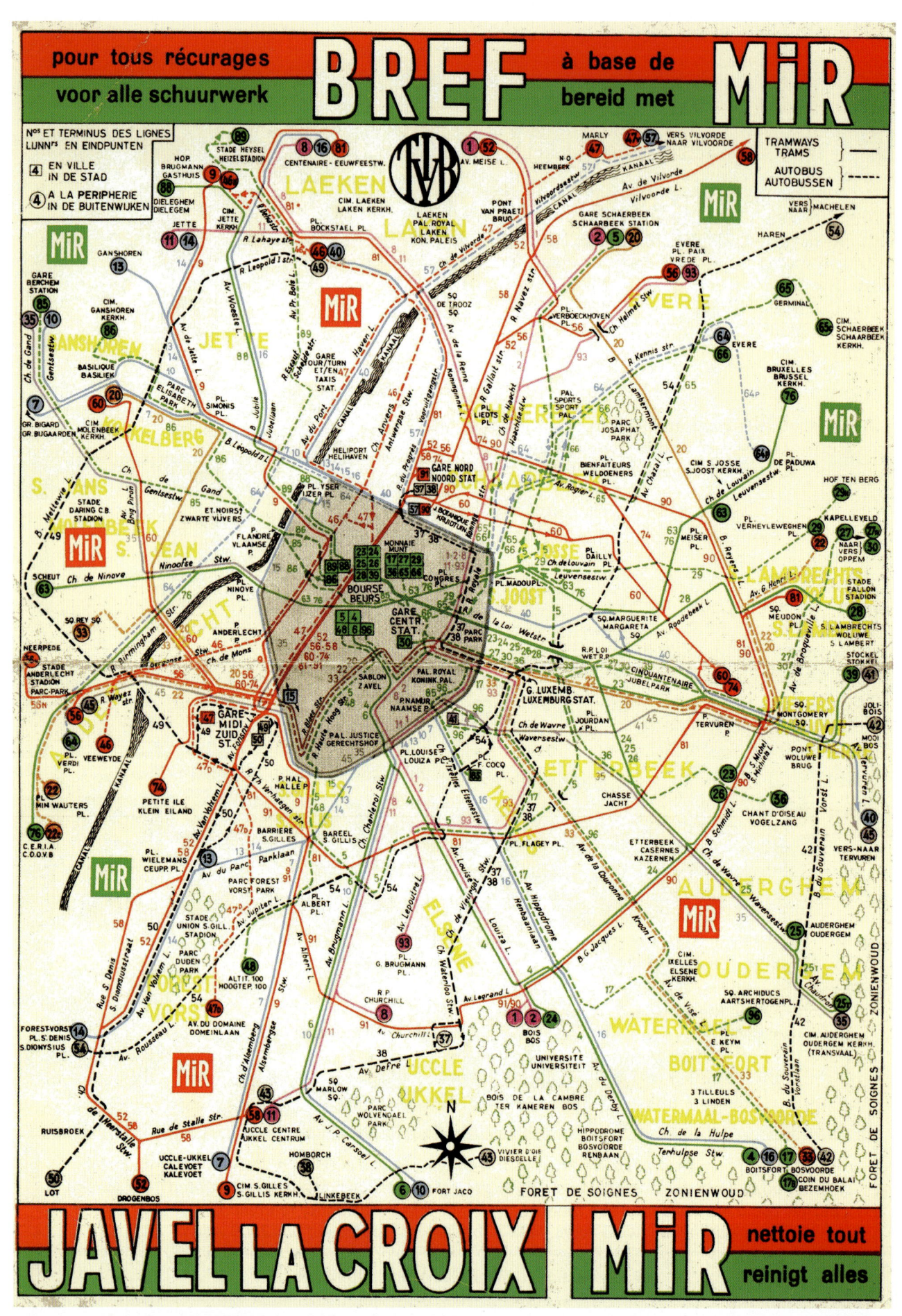

Map 34. When planning began on what became pre-metro line 1 the tramway network was largely complete (only about 30 route-km had closed) and there were over 700 motor trams in service, and 400 buses. The official route diagram for 1962-63 showed a network ready for the start of tidying-up and concentration implicit in the pre-metro plans. [Courtesy Eric Smith]

10.1 A long decline is reversed

Although there was never a substantive decision to abandon tramway operation in Brussels the process of decay and petrification seemed to point in that direction by the 1980s. There was no line extension to the system between 1952 and 1982, no new rolling stock between 1978 and 1993. Although the decline in the extent of the network is partly attributable to the conversion of two pre-metro sections to Metro, which we have described, there was also a steady process of replacement of individual lines by buses. Route length shrank from 153km in 1977 to 131km in 2001. Over the same period the fleet was reduced from 467 units to 292, although the capacity of the latter was greater.

From the 1990s there was a sea change in attitudes to the tramway, partly in response to international trends, partly to the localisation of responsibility for transport to the Regional Government in Brussels, which had different priorities and a need to contain capital expenditure. By 2008 there was not only the arrival of the first of a new large fleet of low-floor trams but also a series of line extensions, with planning for more in the pipeline.

Then consider the performance of the undertaking. Overall traffic on buses, trams, and metro has increased by 60 per cent since 2003. There were 400.9 million journeys in 2017 (286.5 million in 2008), and until 2015 traffic had grown for fifteen successive years (a small fall in 2016 was attributable to the effects of the terrorist outrages and temporary closures that year). Tramway traffic (149.1 million journeys in 2017, up from 74.2 million in 2008) has also grown, partly a consequence of service extension, better and more frequent services, and increased capacity resulting from larger vehicles. Traffic congestion, parking difficulties, and other external factors have also had their effects. This now comfortably exceeds bus traffic (100.2 million journeys), which it roughly equalled in 2008. Tram traffic has returned to a level last seen in the 1960s when the network was, of course, much larger. Most startlingly, total tram traffic is edging toward parity with the total for the Metro (151.7 million journeys), a situation unimaginable twenty years ago.

10.2 Rolling stock update

The most significant event in recent years was the final withdrawal of the last of the 172-strong 7000-type four-axle PCCs, one of the most distinctive and successful of all modern tram designs, associated with the early years of the pre-metro. The remaining cars had been refurbished but internally were still relatively austere as well as inevitably retaining their awkward entrance arrangements. Latterly they served faultlessly the feeder lines linking Ban Eik and Tervuren with Montgomery, providing a Rolls-Royce service for the residents of the affluent suburbs to the east of the city. The cars were finally withdrawn from normal service in February 2010, and amongst them was No 7008 which had operated continuously since

The traditional Brussels tramway just before the revolution of pre-metro construction: at Place Rogier different types of STIB trams (right) pass metre-gauge SNCV trams (left). The Tour Martini and theatre behind the trams replaced the Nord station (and have since been replaced again). [MJR]

Between 1967 and 1988 tram services were constantly affected by pre-metro and metro construction works. A southbound PCC passes the site on the Petite Ceinture which would later become Hotel des Monnaies station. [MJR]

The long-delayed southwards extension of service 51 from Silence to Van Haelen at last opened in September 2008.

The long 7900-type cars benefit from the new livery. This is the recent Marius Renard terminus.

The Chaussée de Charleroi services were upgraded and equipped with new 2000-type trams from 1994 in a successful demonstration project to show the potential of new tramways. No 2005 nears Vanderkindere in 2008, although the impetus has now rather gone out of these routes.

The next upheaval came with the introduction of 3000- and 4000-type low floor fleets, and the reordering of the north-south axis services, including the new transfer arrangements seen here at Rond Point Winston Churchill.

January 1952, a record unlikely to be matched by their successors.

The 127 six-axle articulated cars which followed the first PCCs were equally successful, and by improving productivity helped to secure the system's future. The arrival of the 3000- and 4000-series first enabled disposal of the last four-axle cars but the six-axle cars followed, and by 2016 only 65 remained in use. Thoroughly refurbished the cars continued to provide excellent service but their days are numbered. The 7900 eight-axle series all remain in use, despite their high current consumption, because of their reliability and high capacity.

The first fleet of 51 low-floor cars (type 2000) are problematic. They were a resourceful early attempt to design an all-low-floor car, and emerged from several years of painstaking research and trials in the vain hope of capturing part of the growing market for trams of this type. They were also amongst the last European trams to derive from a specifically 'national' development programme. From the start their unique truck design has given rise to excessive vibration which has limited their effective use, and although still in service their long-term future must be uncertain.

The 3000- and 4000-series Flexity cars now dominate the tunnel routes and other parts of the network. More will follow, leading to the withdrawal of all remaining 'heritage' cars, and in February 2018 STIB announced a prospective order for up to further 175 cars from Bombardier, for delivery from 2020.

In 2017 the number of cars available for use was as follows:

Type	Number available
7700	65
7900	61
2000	51
3000	150
4000	70
Total	397

301 trams were required for daily service during the winter timetable 2017-18.

The current fleet is the largest since the early 1980s, bearing in mind also that the unit capacity is far greater on average than it was then.

10.3 Network

The 2017 tramway system is shown on map 33. A programme of route extensions has continued in recent years, though sometimes slowly. The 2.2km extension between Herrmann-Debroux and Pont de Woluwe, Musée du Tram (with three new intermediate stops) opened to the public on 14th March 2011, over a section previously abandoned in 1963. A second new route, between Place Meiser and Bordet (also with three new intermediate stops), runs mainly along an existing central reservation in Boulevard Léopold III, and was brought in use from August 2010 for depot workings, and opened for normal public service on 1st September 2011. This route, and the Woluwe extension, have been partly equipped with catenary overhead, the first on the system since abandonment of the Erasme route in 1999. From 30th August 2010 the terminus at Bordet (service 55) was altered and cars were extended by one additional stop to Da Vinci. The next extension projected this line further east to the new NATO Headquarters and 'Eurocontrol'. The total route length of the tramway in 2016 was 141.1km, of which 12.1km were in tunnel, 68.5km on segregated surface track, and 48.5km on street.

Work is proceeding on further extensions: a 3.3km line north from Simonis will serve the University Hospital at Jette, reviving the old line number '9'; alterations to Simonis subway station were proceeding in 2017 to accommodate the inner end of this new service, with a three-track terminal siding as well as improvements to the existing station, easing interchange between tram and metro. The new branch is expected to open in September 2018. Also in progress is an extension north from Pont de Woluwe towards Roodebeek for service 94: this too is expected to open in September 2018 and further extension is planned. Longer-term plans include additional lines serving the Heysel plateau, linking Esplanade and Stade termini, and the possibility of a route to serve the massive developments on the Tour & Taxis site west of Nord. The neighbouring Flemish Region has aspirations for outer-suburban lines to the north-west of the city, in effect succeeding some of the former Vicinal lines in the area, closed in 1978, although consideration is also being given to 'bus rapid transit' in the northern sector. Planned in 2017 was a reordering of the fascinating terminus at Montgomery for services 39 and 44, to eliminate the loop with its two single-track portals made necessary by the former use of single-ended cars. The awkward inwards track, already avoided in the evenings to limit noise and disturbance, would be replaced by a double-track ramp on the site of the present outwards portal, and a stub subway terminus.

A glance at map 33 will show how the somewhat disjointed tram network left after the 1970s closures is developing into something more coherent, with new orbital links and an emphasis on intermodal connections, such as the extension of service 25 to the future RER station at Boondael where there is a new terminal siding. A better interchange with the railway at Berchem station (service 82) has recently been completed.

A larger tram fleet, and far larger cars, has placed impossible pressure on the existing depot accommodation, most of which dates from the early part of the last century. The new Haren Depot became fully operational from 6th April 2009, with its own car allocation and the run-out 'index' number '3'. A long-planned new depot on the Marconi site, close

Services along the fine Boulevard du Souverain between Boitsfort and Woluwe were withdrawn as long ago as 1963, long before suburban employment growth and the need for peripheral links became apparent. The relaid line, which took over a traffic lane in each direction, is seen in March 2011.

Extended to Woluwe (Musée du Tram) in March 2011, the Souverain line will be further projected towards Roodebeek. A primary school party enjoys a tram ride in May 2011.

The Bordet line was extended north-eastwards, first to Da Vinci in 2010 where a 2000-type is waiting departure on a wet day.

Beyond Da Vinci the line eventually reached a new terminus at NATO and Eurocontrol, although at present traffic is confined to business hours and the service operates for only part of the day.

Improvements to the system are not limited to extensions, and minor additions have included these terminal sidings at Poelart close to Louise station.

The withdrawal of the last single-ended cars in 2010 removed the need for wye-reversers which had once been characteristic of most STIB termini. At Ban Eik a new stub terminus was installed (and the 7700-type six axle cars replaced the four-axle PCCs).

to Parking Stalle terminus in Uccle, was completed in 2017. This not only provides more suitable accommodation for the new cars but also improves operational efficiency by simplifying depot workings to and from the southern termini.

Financial restrictions limited large-scale renewals for several years and in the 1980s parts of the track and paving were in poor condition, the effects partly disguised by the remarkable riding qualities of PCC-type bogies. A major programme of renewals, including relaying of complete lines, followed in the 1990s and continues to the present day, with large-scale programmes each year to renew major intersections, such as those at Uccle (Globe) in 2012 and Buyl in 2017. 12.1 single-track km were replaced in 2012.

An important development for the long term is the gradual respacing of tracks when renewals are undertaken, allowing eventually the operation of wider and more spacious vehicles.

10.4 A look ahead

As we look towards the future it is clear, as it was not forty years ago, that the tramway is permanently established as a major -if secondary- mode in the Capital Region. Investment in rolling stock, track, depots, and plant, as well as a series of extensions, make this sure. Of course there are uncertainties. The future politics of the Belgian state, and hence of decision-making and funding, may become significant. Tourism may be adversely affected by recession or terrorism. Automation and artificial intelligence may profoundly change the nature of work, and autonomous vehicles and systemic ride-sharing may affect demand for public transport. Employment, especially in relation to the EU and related institutions and enterprises, may suffer with effects on traffic and revenue: although small in relation to overall numbers the departure of UK parliamentarians, their staffs, and British employees of the Commission is a symbolic and adverse change following the nation's decision to leave the EU.

When line 3 becomes part of the Metro system in the late 2020s it will have a profound effect on the shape and nature of the tramway. Cross-city north-south running via Rue Royale will of course continue, but the nature of the streets served limits the efficiency and capacity of this corridor. On the line 3 north-south axis the present pattern of direct connections, already reduced, will change again. From Rogier and Gare du Nord feeder tram services would continue towards Esplanade and Meiser, but tracks along the present route of service 55 will be abandoned. Ironically those who at present enjoy through services to and from line 3 will lose them, whilst those resident along part of service 55 will gain them (passengers from east of Bordet will have to change). At the southern end feeder tram services will continue from the new Albert terminal towards Vanderkindere and beyond, and to Uccle, and service 51 at present running through from Heysel will

Although several routes now benefit from the highest standards of modern light rail practice there are still places where narrow streets interrupt free movement. One is Avenue Legrand near Bois de la Cambre on the prestigious east side peripheral line incorporating pre-metro section 5. This is used by key services 23 and 24 and there have been proposals for over half a century for remedial measures here. Cost and practicability have so far prevented action.

The nearby Jamar ramp was reduced from four tracks to two in connection with the construction of the tram link into the deep-level Midi station, finished in 1993 (see page 3).

New depots have opened at Haren (2009) and Marconi (2017), and there have been improvements at the older sites, notably at Ixelles which is responsible for routine repairs to the modern fleet. A 2000-type receives attention in 2013.

Track renewal is a continuous process, tending now to be undertaken during extensive closures for a period of weeks. This was the busy crossing at Uccle (Stalle).

Back where we started: in 2017 much of the Midi subway remains in use, and this is the Poincaré ramp as it is today, having sadly lost much of its original art deco style (see page 33).

To the future: in 2017 work was proceeding well on building the new line between Simonis and Haut de Jette, following a similar alignment to former service 9 which closed in 1968. More extensions will follow. [Courtesy Christian Scheemaekers]

necessarily be truncated. Details of the actual pattern of future tram operation are awaited.

No amount of public relations gloss will alter the fact that for many suburban travellers journeys will become less convenient. It remains to be seen how well-arranged the interchanges are, and how crowded they become.

To an extent the new tramway network will have taken on the form envisaged for it fifty years ago: a series of high-frequency feeder lines linking with Metro services; and peripheral and inter-suburban lines catering for journeys away from the centre where the volume of traffic is unlikely to merit the investment needed for Metro construction: the east side peripheral line (5) is already a classic, and largely effective, example of modern light rail and could be further improved relatively cheaply. Further tram extensions are likely, not least because of the availability of rolling stock made surplus from line 3, and several are planned. It may also be hoped, surely, that the long-delayed measures to deal with surface traffic congestion at such points as Meiser and Buyl will be at last addressed.

It would be wrong to paint too rosy a picture of the Brussels tramway which inevitably suffers, in comparison with the new French systems for example, from the consequences of its traditional heritage in terms of depot location and accommodation, track layout, curves, and other restrictions. Despite the introduction of more public transport priority schemes, scheduled speed on the tramway remains stubbornly low, at around 16 km/h compared with 28km/h for the Metro. In Köln, with a significantly larger spread of segregated surface track, the average is about 26km/h, showing what could be achieved. Short of more underpasses and segregation –very costly to achieve-- it is hard to see what further measures could be taken to improve this in Brussels (the eventual conversion of line 3 will have an adverse statistical impact on the apparent performance of the remaining tramways).

10.5 Tramway services in 2017

(Services in **bold type** use pre-metro lines 3 and 5) Source: *https://www.stib-mivb.be/?l=en* (August 2017)

3	**Esplanade — Gare du Nord – north/south tunnel – Gare du Midi – RP Winston Churchill (13.2km) ***
4	**Gare du Nord – north/south tunnel - Gare du Midi – Vanderkindere – Parking Stalle (9.7km) ***
7	**Heysel – De Wand – Meiser – Montgomery – Legrand – Vanderkindere (16.1km) ***
19	Groot Bijgaarden - De Wand via Simonis (inter-suburban service) (9.2km)
25	**Rogier – Meiser – Boondael Gare (11.4km)**
32	***Da Vinci – north/south tunnel – Gare du Midi - Forest – Drogenbos** (after 20h00 only; see also line 82) (15.3km)*
39	Montgomery - Ban Eik (8.8km)
44	Montgomery – Tervuren Station (9.4km)
51	Stade – Gare du Midi – Uccle (Globe) – Van Haelen (15.5km)
55	Da Vinci – Gare du Nord – Rogier (5.9km)
62	Eurocontrol - Da Vinci – Meiser – Cimetière de Jette (10.3km) *(part day between Eurocontrol and Da Vinci; not Saturdays, Sundays or public holidays)*
81	Marius Renard –Gare du Midi – Montgomery (12.1km)
82	Berchem - Porte de Ninove –Gare du Midi – Drogenbos (13.4km) *(Operates Berchem – Midi only after 20h00: see also line 32)*
92	Schaerbeek – Rue Royale – Chaussée de Charleroi -Fort Jaco (12.6km)
93	Stade (Houba de Strooper) – Rue Royale – Avenue Louise – Legrand (11.5km)
94	Place Louise – Herrmann-Debroux – Musée du Tram (11.8km)
97	Place Louise – Barrière - Forest – Uccle – Dieweg (9.5km)

Source: *https://www.stib-mivb.be/?l=en* (August 2017) **Lines branded 'Chrono', i.e. tram services equivalent to Metro performance in terms of reliability, speed, and rolling stock standards.*

11. Pre-metro: delusion or salvation?

"L'avenir n'est plus ce qu'il était."
("The trouble with our times is that the future is not what it used to be.")
(Paul Valéry, 1871 - 1945).

The French model: pedestrianisation and traffic management leave a 'free way' for the trams. The highly distinctive cars in Marseille glide through the city centre, mostly on the surface.

121

In our story of transport in Brussels, Antwerpen, and Charleroi we moved from the bleak landscape of ravaged post-war Europe to the different, but in some respects equally uncertain, present day. The tomes of plans we have glimpsed along our way were mainly discarded, and what actually occurred on (and below) the ground was only a tiny part of what was hoped for. But there was an outcome, as we have seen, and much good has come from what was intended for so long, even if some of it is not quite what was expected.

Looking back over what appears to be a desert of troubles and disappointments we seem to find in the prosperous, optimistic 'sixties, especially, an oasis of optimism and determination. It wasn't quite like that, of course, but in some fields at least there then seemed to be certainty and ambition where now there is often doubt and division. The plans for radical renewal of the Brussels tramways belong with contemporary aspirations for social housing, architectural innovation, town planning, new employment opportunities, and much more, brought to a juddering standstill by the oil crisis of 1973-4 and the troubled austerity which followed (and which is still with us, in different guise). It is fashionable now to despise those days, anathema as they are to neo-liberals and post-modernists, but it will have been clear that the authors don't wholly share those attitudes.

The pre-metro concept, as it evolved more in Antwerpen and in cities such as Stuttgart and Köln than in Brussels, promised a great deal. It was an economical, cumulative project which would gradually bring the advantages of smoother, quicker travel to more people whilst preserving established and comprehensive transport links. But there was a downside too, partly avoidable but not always avoided. The two-stage process could be wasteful of resources, with expenditure on aspects of the interim phase discarded prematurely when conversion occurred. It was this alleged profligacy which helped to condemn line 2 in Brussels in 1988.

There was also a certain degree of irresolution in Belgium even before the financial restraints began. For instance, the continued long-term use of the 1957 tunnels at Midi for line 3 trams was a long-extended impediment which could have been addressed: frequency on the whole line is still inhibited by the flat junctions within the old tunnel. Delays in commissioning near-complete infrastructure in Antwerpen and Charleroi reached almost farcical levels, as we have seen. Other features on the traditional Brussels tramway were allowed to continue because there was no essential need for change, a contrast with the circumstances of the 'new' tramways which began with a clean slate. Only in the new century did these and other matters begin to be systematically addressed in Brussels (and more sections in Antwerpen and Charleroi were completed). There is an argument, unpopular needless to say amongst 'enthusiasts', that a fresh start was not always a bad thing. The STIB's Annual

In Stuttgart, as we have seen, an almost perfect pre-metro concept developed, with city centre subways and high quality surface extensions. But at a price in terms of capital and disruption, and not necessarily always convenient.

Ludwigshafen was one of the German cities which built underpasses, including this one at the Hauptbahnhof. In this case, although the tracks remain for service use, the line is no longer in public use.

Rouen has a rare modern example of a lengthy downtown subway.

Bremen has a successful and extensive tramway system without subways, and was a pioneer of city centre pedestrianisation.

The surface tramways across the Frankfurt-am-Main old city became a cause célèbre, and their salvation after court intervention was an important turning point.

Some of the new French tramways have incorporated underpasses where necessary, like this one in Montpellier.

Reports regularly bewail its performance compared with that of the 'new' French tramways.

Reynaert also underestimated the immense difficulties inherent in constructing, in an old and constricted city fabric, connecting surface lines able to offer the same high standards of reliability and efficiency as that provided by the new tunnels. The cost of land acquisition, and in more recent times opposition on amenity grounds to extensive clearance, prevented anything like the extensive network of new tracks which was intended. In Germany, it is instructive to see, surface track improvement moved hand-in-hand with tunnelling, creating in several cities a seamless rapid transit system akin to a U-Bahn in its effectiveness but at lower cost.

Private road traffic, meanwhile, continues to grow in volume, and congestion, solved in one place, soon moves to another. As a result tramway operating speed continues to be a nagging difficulty across the Brussels undertaking and there are still too many places prone to chronic delay. Lastly, there was the unsuitability of existing rolling stock, a major factor on line 1 in Brussels, and to an extent in Antwerpen. Solutions to that problem already existed, and nearly a decade later were proposed, but in vain. Only with the unanticipated debut of the low-floor car was the problem eventually addressed.

When the tide somewhat unexpectedly turned, in France, Spain, and Great Britain notably, the revival of tramways proceeded along simpler, more easily achievable lines than the grander plans the 'sixties had envisaged. New sub-surface operation is now largely confined to minor underpasses, and outside the United States comprehensive new tram subway installations are almost unknown. Karlsruhe offers a rare recent exception, where excavation has proceeded for several years on a first central area tunnel, said to be necessary because of a great increase in tramway traffic following the expansion of the 'tram-train' network. Instead pedestrianisation, the zone *piétonne*, became the key which unlocked tram access to city centres. Given a choice the general public probably prefer to stay on the surface, and the gently moving trams add to the vibrancy of the city scene as well as permanently advertising the 'transit presence'.

Amongst the earliest examples of pedestrianisation was the area around the main station in Zurich, but they have become almost commonplace around the world, in Melbourne for example as well as in Europe. It is interesting to note that one of the principal tram subways planned to avoid traffic in Leeds in 1944 would now have lain below a main street (Briggate) which is now wholly car-free. Within Brussels a rather tentative recent pedestrianisation project embraces the area around Boulevard Anspach, once a notorious spot for conflict between trams and traffic: the tram is now far below ground and the road traffic has largely vanished. Gent, eschewing tram subways which would have been an act of barbarism

The new Scottish tramway, in the capital Edinburgh, shows the advantages of traffic management and careful track layout: St Andrew Square, July 2017. [Courtesy Martin Dibbs]

there, has skillfully crafted a city centre pedestrian zone through which trams now glide, more slowly it is true, but taking into account the wider-spaced underground stops and the time taken to transit between surface and platform, the overall travel time is little different. A decisive moment in the move against extensive tram or metro subways occurred in Frankfurt-am-Main in 1986. With the continued spread of underground lines there it was proposed to discontinue the last surface operation across the old city. In an unexpected public response 60,000 petitioners succeeded in appealing the matter to the Frankfurt *Landgericht* (Court), which in a landmark decision ordered retention of the existing tramway. It continues to run today with increased frequency. Here and elsewhere a radical change is that street running in general traffic is now again an accepted technique, although with trams given priority and careful traffic management.

France had shown the way with a characteristically different approach. The urban tram had disappeared almost as completely as it did from Great Britain, and from the 1960s only three attenuated undertakings survived, in Marseille, St-Etienne, and Lille. But from 1975 a drastic reappraisal of national transport policy transformed the situation, owing much to the dynamism and persistence of M. Marcel Cavaillé, France's long-serving Transport Minister from 1974, who wrote to city Mayors urging adoption of innovative transport systems, funded partly by hypothecated taxes. Notable as part of these transport changes have been the successful efforts to 'insert' the new trams into a revived and harmonious city scene as part of wider projects of urban renewal.

Traffic management, freeing up road space, has been another factor leading to allocation of streets to public transport only, as in Manchester. Here and elsewhere an unforeseen development which has been critical in building new tramways has been the availability of former railway rights of way for adaptation. It seems likely that, in Europe at least, future applications of tram subways will be of limited extent: the new Granada system in Spain, for instance, has just three of its 26 stops underground, and subways elsewhere have been largely confined to locations where special conditions require them; in France Rouen and Lille are examples, as is the restored tunnel in Marseille. In the United States there are notable examples in, amongst other cities, Los Angeles and San Francisco (where the major 'Central Subway' is under construction in classic style). Elsewhere limited underpasses have been the preferred option, as in Montpellier, Manchester, and Sheffield, a reversion in a sense to the earliest days of subway construction.

In West Germany the availability of generous central government funding over a long period of years encouraged public transport improvement on a scale unequalled elsewhere, although funds were not always perhaps wisely spent. Eventually fifteen cities had tram subways, with minor underpasses as well. Notably the more straitened finances of Eastern Germany (the 'DDR') did not allow such elaborations, and of course with private motoring so restricted the need was less. A classic central subway network was planned for Leipzig but never built. Some German cities launched full U-Bahn operations from the start: München (where they had been planned before 1945), and Nürnberg were added to the older installations in Berlin and Hamburg, which were extended. Some large German cities operated successful tramways without subways, notably

The most sensitive areas can be traversed by well-designed surface tramways, as in Orleans where wire-free surface current collection has been installed near the Cathedral. Also apparent here is the French practice of thoroughly overhauling the whole urban townscape when building a tramway.

Bremen, Kassel, and, until recently, Karlsruhe and Dusseldorf. Unlike in Brussels none of the German pre-metro (or 'U-Strassenbahn') installations has been upgraded, even if they have been rebranded for marketing purposes.

So, was the pre-metro a misguided project, an unnecessary way-point on the journey towards full Metro, or a costly way to avoid removal of trams altogether? We have seen the notable contrast between Brussels and Antwerpen, confused as it is by the economies which delayed completion by decades, but now showing in the latter city the merits, as Reynaert rightly foresaw them, of a comprehensive net of through services linked to a high-quality central core, the best of both worlds. In Brussels, even if long delayed by indecision, line 3 has for over a decade now successfully delivered many of the benefits of metro operation but with a wider reach (which will be diminished by its eventual conversion). The 'pre-metro tunnel-plus-tramway-reservation' solution—could, if properly resourced and operated, extend the fruits of segregated running more widely and more cheaply. Simply put, you get more for less money. It is at least arguable that, if properly equipped, line 2 in Brussels could have flourished under pre-metro conditions and its services would have reached further into the suburbs as a result, as we have seen. This is what has developed in Antwerpen, where thoughts of conversion long ago disappeared.

The classic Metro has a crucial part to play, where conditions and traffic volumes are favourable. But its

This scene shows what will be lost when line 55 in Brussels is put underground. This Flexity tram is calling at Waelhem stop inbound through traffic-free streets to the city centre. Surely there could be no closer integration between a high-quality, clean transport system and the community it serves? Intervisibility between tram passengers and local inhabitants couldn't be more complete, enhancing security, promoting high-class affordable transport, and enriching the image of the city and neighbourhood, all at the same time. Why bury the trams? [Courtesy David Holt]

characteristics can make it a clumsier technology, its stops expensive to build and wider-spaced, its trains suitable for the densest traffic flows, its tunnels not always attractive to the user. By contrast the tram is often seen as a more human system, and allied to short subways -where absolutely necessary- it can offer almost equal advantages.

So we should not see the pre-metro concept as a transitory expedient, seeming to be appropriate in its day, but as a project which delivered many of the results it promised, and which remains a valued legacy of an era in which investment in infrastructure was not seen as a begrudged necessity but as a down-payment on the future public good.

12. Acknowledgements

In the course of our long studies of the Brussels and Antwerpen systems we have been much helped by comments and information from friends and correspondents, and we would like especially to thank the following: the late Richard Buckley, Martin Dibbs, Tim Figures, Christoph Heuer, David Holt, Roger Jones, Anja De Neve, Joachim Nijs, Brian Patton, Mike Russell, Eric Smith, Michael Taplin, Mark van den Eynde, and David Verrezen. Mike Willsher, Honorary Librarian of the LRTA, helped us greatly by locating early articles. We are grateful to Andrew Braddock, Carl Isgar, and the LRTA Publications Committee for their support. Thanks are finally due to the Belgian associations *MUPDOFER* and *Tram 2000* for invaluable information over many years.

The diagrams, text, and illustrations were originated over a period of seven months on a Macintosh i-Mac 2.7 GHz Intel Core i5, with Epson V550 scanning equipment and Adobe Photoshop processing software. Its full use was again made possible by the generous and expert assistance of Steve Xerri.

Most of the illustrations are by the authors or from their collections, but several generous individuals kindly made their pictures available. In particular we could not have published the book without access to the astonishing resources of Michael J. Russell. Our thanks for permission to reproduce illustrations also go to John Bromley, Martin Dibbs, Tim Figures, David Holt, Tony Percival, Christian Scheemaekers, Patrick Sellar, and Michael Taplin. Photographs by the late Jack Wyse and the late Frank Hunt are reproduced by courtesy of LRTA Collection/OTA (http://*www.onlinetransportarchive.org*), with thanks to Roger Jones and Peter Waller. Photographs from the collection of the National Tramway Museum are reproduced by courtesy of the Tramway Museum Society.

Many of the maps and diagrams, adding so much to the qualities of the book, were drawn specially by Roger Smith, doyen of tramway cartographers. *MUPDOFER* and *Tram 2000* have helped us immensely. Geoffrey is especially grateful to David Hall, whose support has made this book possible.

12.2 Select Bibliography

The City
The break-neck redevelopment of parts of Brussels is essential background to our story, and is described *inter alia* in the following works:
Pierre Mardaga *Commerce et Négoce* (Bruxelles 2003).
Direction des Monuments et Sites de la Region de *Bruxelles-Capitale* [series *Bruxelles, Ville d'Art et d'Histoire*].
    5 *Le Heysel.*
    20 *Les Boulevards du Centre.*
    40 *Les Boulevards Extérieures.*
    44 *La Cité administrative de l'Etat.*
Anne-Marie Bogaert-Damin Bruxelles : *Développement de l'ensemble Urbain, 1846-1961: analyse historique et statistique des recensements* (Namur 1978).
Patrick Burniat et al *L'Architecture Moderne à Bruxelles* (Louvain-la-Neuve 2000).
Pierre-Yves Monette *Belgique, où vas-tu?* (Wavre, Belgique 2007).
Charles Picqué *Pour Bruxelles: entre périls et espoirs* (Bruxelles 1999).

Tramways and Subways
Nils Carl Aspenberg *The Tramways in Stockholm* (Oslo 1998).
Bradley H. Clarke and O.R. Cummings *Tremont Street Subway, A Century of Public Service* (Boston 1997).
F. de le Court *La Jonction Nord – Midi: son Caractère Hydraulique et le Problème des Égouts in Science et Technique,* Nos 6 – 9, November 1949 – April 1950.
Harold E. Cox *The Road From Upper Darby* (New York 1967) [Philadelphia subways].
W.J.K. Davies *The Vicinal Story 1885-1991* (Scarborough 2006).
Mike Filey *The TTC Story: The First Seventy-five Years* (Toronto 1996).
Paul Garbutt *World Metro Systems* (Harrow Weald 1989).
Christoph Groneck and Dirk Martin Stein *Metros in Belgien* (Berlin 2009).
David R. Keenan *The North Sydney Lines* (Sydney NSW 1987) [Wynyard subway].
Luc Koenot et al
--: *Flash 1996 Belgique* (Third edition, Bruxelles 1995).
- -: *Flash 2006 Atlas des Tramways, Metros, et Trolleybus Belges* (Fourth edition, Bruxelles 2006).
Alex Krakowsky *De Antwerpse Tram-r-evolutie* (supplement to Tramfan journal, 2002).
M. Lombard *Le Viaduct Sud in Revue Trains,* No 13 December 1947.
Jean-Pierre Marissens (ed) *Historique des Lignes des Tramways Bruxellois* (Bruxelles 2002).
Herman Mulkay
--: *Les Tramways aux Abords de la Nouvelle Gare du Nord in Revue Trains*, No 2 June 1951.
- -: *La Circulation aux Abords de la Nouvelle Gare du Midi in Revue Trains No 1* (new series).

MUPDOFER *TB/TUAB/STIB Historique du Matériel Roulant* (Bruxelles 1974).
Brian Patton *The Development of the Modern Tram* (Brora n.d.) [especially on 'hybrid' cars].
Geoffrey Skelsey & Yves-Laurent Hansart
---: *Brussels: A Tramway Reborn* (LRTA 2008).
---: *PCCs of Western Europe 1950 – 2010* (LRTA 2011).
---: *Charleroi's Trams Since 1940* (LRTA 2013).
Douglas N.W. Smith *Canada's First Subway: From Conception to Operation* (in 3 *Canadian Rail Passenger Review* (Ottawa July 2000), pp. 90-94).
Société des Transports Intercommunaux de Bruxelles
---: *'Metro? Oui!'* (Bruxelles 1968).
---: *STIB 2020: Visions d'Avenir pour le Transport Public Urbain à Bruxelles* (Bruxelles 2004).
---: *Metrovision (Bruxelles 2009)*.
Wiener Stadtwerke *Eine U-Bahn für Wien: Planungstand 1966* (Wien 1966).

Periodicals
Historail
Modern Tramway
Modern Railways
Nos Vicinaux
Rail & Traction
Railway Gazette International
Tramways & Urban Transit
Tram 2000

Le Soir
La Libre Belgique

Annual Reports of the TB and STIB 1945-2016

For current maps and timetable information of all STIB modes see the website:
https://www.stib-mivb.be/?l=en
The site also links to factual and statistical information covering several years, including Annual Reports.

Information on operation in Charleroi is available at *https://www.infotec.be*;
and in Antwerpen at *https://www.delijn.be/nl/lijnen/*.

Particular thanks are due to the Cambridge University Library and to the London Guildhall Library for research facilities. The sadly now closed Brussels bookshop *Libris* in Espace Louise, which in its former home nearby had a subway portal almost outside the door, was for many years our invaluable source of lesser-known books and pamphlets on the city's history, and we pay tribute to it.

About the Authors
Geoffrey Skelsey first visited Brussels as a schoolboy in 1959, and fell in love with the then nearly complete tramway, as well as the distinctive character of the city itself which became for some years almost his second home. After graduating from St Catharine's College, Cambridge, his professional career began in the transport industry but he subsequently moved into public administration and later politics as a speech-writer at Westminster. From 1984 he spent increasing time in Brussels, providing the basis for much of this study.
The authors met over twenty years ago, and have since co-operated closely on the study and documentation of Brussels transport in four books and many journal articles. Yves-Laurent Hansart was born and educated 'within the city walls' of Brussels and his interest in trams was fostered by their passage outside his childhood home, and on his daily journeys to school. He is an active participant in the work of the Association Pour la Sauvegarde du Vicinal (ASVi) and lives with his family in Brussels.

The Light Rail Transit Association

Advocating modern tramways and light rail systems

The LRTA is an international organisation dedicated to campaigning for better fixed-track public transport, in particular tramways and light rail. The Association celebrated its 80th anniversary on 30 June 2017.

Membership of the LRTA is open equally to professional organisations, transport planners and individuals with a particular interest in the subject. Members receive free of charge by post *Tramways & Urban Transit,* the Association's all-colour monthly magazine, as part of their subscription. With tramway and light rail systems being adopted not only in Europe but world-wide, this high-quality journal features topical articles and extensive in-depth news coverage as well as trade news and readers' letters. Details of local meetings in the British Isles are also included.

The LRTA also publishes *Tramway Review* – a quarterly journal devoted to historical material.

Officers of the Association – many with transport industry experience – form part of an extensive network of light rail and tramway information sources, which includes the comprehensive LRTA library.

For more information visit our website: **lrta.org**

To become a member of the LRTA go to: **lrta.info/shop** or e-mail **membership@lrta.org**
Postal address: **LRTA Membership, 38 Wolseley Road, SALE, M33 7AU**

For general enquiries contact: **secretary@lrta.org**
Postal address: **LRTA Secretary, 138 Radnor Avenue, WELLING, DA16 2BY**

To order copies of our wide range of books go to: **lrta.info/shop**
Orders may be sent by post to:
LRTA Publications, 31 Ashton Road, WOKINGHAM, RG41 1HL

Books due to be published by the LRTA over the next eighteen months include:

Modern Trams - Volume 1

The Tramways of Portugal

Tramways and Stadtbahn in Hannover

Belgium Underground – Pre-Metro and Metro, 1957 – 2017

Tramways in Bolton

Great City Tramway Albums – Vol 1 Vienna

The Leaving of Liverpool – The Story of Liverpool trams 1945 - 1957

Potential authors of books on subjects relevant to the Association's interests are invited to contact the Chairman of the LRTA Publications Group at:
24 Heath Farm Road, FERNDOWN, BH22 8JW